Your Cat's
First Year

Your Cat's First Year

Photographs by Jane Burton and Kim Taylor
Text by Michael Allaby

Simon and Schuster New York

Published by Simon and Schuster
A Division of Simon & Schuster, Inc.
Simon & Schuster Building
Rockefeller Center
1230 Avenue of the Americas
New York, New York 10020

First published in Great Britain in 1985
under the title *Nine Lives* by Ebury Press

SIMON AND SCHUSTER and colophon are registered
trademarks of Simon & Schuster, Inc.

Designed by Gill Della Casa

1   2   3   4   5   6   7   8   9   10

Library of Congress Cataloging in Publication Data

Burton, Jane.
    Your Cat's First Year

    Includes index.
    1. Cats. 2. Cats – Development. 3. Cats – Behavior.
I. Taylor, Kim. II. Allaby, Michael. III. Title.
SF 442.B89 1985   599.74′428   85-10743
ISBN 0-671-60090-7

This book was produced in association with
Bruce Coleman Limited

AN EDDISON • SADD EDITION
Edited, designed and produced by
Eddison/Sadd Editions Limited
2 Kendall Place, London W1H 3AH

Phototypeset by Bookworm Typesetting, Manchester, England
Origination by Columbia Offset, Singapore
Printed and bound by Tonsa, San Sebastian, Spain

Contents

The parents

Snorkel

Fergus

The kittens

Barley

Tabitha

Ferdinand

Fred

Ginger

Lucifer

Septimus

Introduction

I started taking in 'rescue' cats almost by accident. I happened to have a few empty animal pens and so offered to foster one or two cats while new owners were found for them. The one or two soon became an avalanche, especially in summer, the season of moggy kittens.

Snorkel and Fergus came to me as rescue cats during one such busy period. Fergus was a stray and Snorkel a farm cat who had lived all her life in a Domesday barn in Dorset. She came to me when the farmer retired and her home was given to the National Trust. With so many kittens available it was impossible to find homes for middle-aged Snorkel and Fergus then – so they stayed on with us.

When Fergus arrived he of course had no name and after some weeks, I simply found myself calling him *Fergus.* Snorkel was so called because an obstruction in her nose caused her to snore nearly all the time, and especially loudly when she purred. Strangely, while giving birth she hardly snored at all, except when purring contentedly between kittens.

I had not originally intended to photograph Snorkel's kittens being born. I had already photographed other cat births and felt sure Snorkel would kitten inconveniently at night. But the kittens started to appear midmorning, so with their birth this book was born.

At first we planned to pick just one kitten for stardom, but soon all the kittens were revealing their different talents and we decided to record not one but the family as a whole, a galaxy rather than one star.

Recording the development of the kittens was my province. Kim Taylor, my husband, concentrated on cat movement: he designed, built and perfected specialized high-speed flash equipment for this project. Both branches of our photography often proved to be two-person jobs, so the pictures in this book are the result of a true team effort by photographer and photographer's assistant (with the roles reversible), and the nine stars. Michael Allaby has completed the book by contributing the main text. He not only describes with clarity the growth and development of the kittens throughout their first year, but also explains many features of their behaviour.

I hope you enjoy reading NINE LIVES as much as we enjoyed producing it.

Jane Burton

Jane Burton
Albury, 1985

The beginning

Our entrance into the world is the most dramatic event in our lives, and the short period that follows it is one of the busiest and one of the most dangerous. There is a great deal that can go wrong.

We forget that time. All animals do, cats no less than humans. While it lasts kittens, like human babies, are driven by the demands of their own bodies and by their own urgent need to control those bodies. In its first week of life a kitten must double its body weight. It must start learning how to use its limbs, how to interpret the signals reaching it through its senses, how to avoid danger by finding its way back to its mother if it realizes it is alone.

The mother cannot leave her kittens during these very early days. She must guard them at all times, and their rapid growth is paid for with food supplied by her body.

Snorkel has been out of the nest, leaving the kittens for a short time, and her return is greeted by demands for food. She settles down as the youngsters climb, burrow and shove in their search for milk. The kittens are eight days old.

Pregnancy

By mid-April it was obvious that Snorkel was pregnant and carrying a large litter too, by the look of her. Repeated child-bearing had made her figure very baggy anyway, but now she really bulged. We did not know when the kittens were due, as we had not been sure of the mating date. Two weeks or more before they were born we were convinced the kittens would arrive at any minute, but Snorkel went on gently swelling. Hers was quite a late litter – other cats we knew of had had their kittens already.

Snorkel about two weeks before the birth of her kittens, showing her pronounced bulge.

Cats rarely suffer any complications during pregnancy. Some weeks after she has mated and before she starts gaining weight and looking obviously pregnant the cat will become less active and her appetite will increase. She will start washing herself much more often than usual, paying particular attention to her abdomen and genital region.

Pregnancy lasts an average of 63 days, but it can vary from about 55 to 70 or more. Domestic cats that spend most of their time indoors tend to carry their kittens for longer than cats used to an outdoor, more active life.

As the pregnancy approaches its term the cat will find a place to make a nest in which the kittens will be born and cared for while they are very young. A cat that lives outdoors will hide. The birth of farm kittens is a secret affair often announced to the farmer by the arrival at his dairy of a mother followed by her alert, playful and hungry family of five-week-olds. House cats feel safer and are less likely to hide. They often choose cupboards or large boxes, but any quiet place may do. Once she has made her choice the cat will spend much of her time in and close to the nest, impregnating it with her smell, which will soon be needed to guide the kittens back to her.

The birth

As the time for the birth draws close the mother's body prepares. Her teats become more prominent and there is swelling in the genital region. These parts of her body are now extremely sensitive and she continues to wash them frequently. The search for a nesting site becomes urgent. In the wild she would choose an old burrow or a hollow tree. It is not only privacy she seeks. The nest must be dry and sheltered from the wind. Kittens can die from cold all too easily.

Her mammae fill as she starts to produce colostrum, the first food the kittens will receive in the world outside her body. Her vulva swells more, until it is distended and starts discharging mucus. She is restless, now squatting down, now scratching the ground, and sometimes calling as she did when she was in oestrus, or growling. Until now her appetite has been good. When she refuses to eat it is a sign that she will go into labour within the next few hours.

It can go wrong

Most cats cope perfectly well with the birth of their kittens. Obviously, cats would have died out long ago were they unable to manage their own reproduction. Yet the same might be said of humans, and we know that among humans there can be complications. There can be complications among cats, too, and the mother may be in distress and need help, or may not have time to do all that is necessary for one kitten before the next appears. It is not unknown among cats living wild for the mother to be attended by another female, usually her sister, who acts as a midwife. The house cat may be quite alone with no one to help her but her human owner. When your cat has her kittens you should be on hand in case you are needed, and you should be ready to summon professional veterinary help in an emergency. It is an emergency if the cat is clearly exhausted or very distressed and has been straining for two hours or more without producing a kitten. A long delay between kittens is not a cause for alarm if the cat seems to be resting and is not distressed.

As with humans, it is possible for a foetus to be presented in the wrong position. Kittens are usually born either head-first or with the hind legs extended and tail-first. If the kitten is in the tail-first position but with its hind legs drawn up beneath its body the birth can be slow and difficult, and if it is in the head-first position but with its head turned to one side veterinary help may be needed.

Snorkel was born and bred as a farm cat. All her previous litters were born well out of the way of humans, among the sacks and straw bales of a barn. This time the event took place in public, yet she was quite unconcerned. Had she been alarmed by the electronic flashes or clatter of the camera Jane would have stopped photographing the event and then there would have been no book!

Diary · Labour

Being a healthy and experienced mother, Snorkel appeared to have no difficulties with the birth of any of the kittens. She also knew precisely what to do with the kittens when they arrived.

By breathing hard and deeply between contractions Snorkel increases the amount of oxygen in her blood to provide more energy. She is working to give birth to her second kitten, Tabitha.

11

Whiskers pushed forward as the muscles of her face strain with the rest of her body, Snorkel pushes hard to expel Tabitha, her second kitten. Barley, the first-born, is already dry.

One at a time the kittens are pushed from the uterus and into the birth canal. As each enters the canal, it remains attached to the placenta by the umbilical cord and completely enshrouded in the membranes of the amniotic sac.

Each of Snorkel's kittens made a brief appearance wrapped in its amniotic sac, was withdrawn into her body, and then reappeared, the membranes broken, and expelled fully. The sac must be ruptured to allow the kitten to breathe. A normal kitten will start breathing without any help, although the first breath can be hard work, with the kitten gulping desperately, almost as though it were drowning. That first breath is taken at once, before the umbilical cord is severed and before the placenta expelled.

Just in case of kitten emergencies it is useful to have a little basic equipment prepared. A roll of cotton wool and some small pieces of clean rough towel may be needed, and you should be able to obtain quickly a large bowl of water at about 100°F.

If the kitten is born still enclosed in the amniotic sac and the mother does not free it at once, help is needed urgently. Holding the kitten on the palm of one hand, even if the placenta has still not appeared, you should use a small piece of clean towel to wipe the membranes and mucus away from the face in general and the nostrils in particular. If the kitten does not breathe at once, rubbing it 'the wrong way' – from tail to head – should encourage it.

Several times Snorkel got up and turned around in the nest dragging a kitten behind her by its umbilical cord. This is common, and mothers often treat their new-born kittens very roughly. They may bounce them along the ground by their umbilical cords, sit, lie or tread on them, push them around with their paws or pick them up and drop them again and if, understandably, this makes them yell, she will take no notice. The kittens are tougher than they look, and they survive.

Some cats lick their kittens at once, moving them forward at the same time and so helping to pull out the placenta. Snorkel took no notice of her offspring until she had expelled its placenta. Then she licked the kitten, bit through the cord and chewed it to within about an inch of the kitten's body, and ate the placenta. Eating the placenta disposes of it in the most economical way, and it also conserves resources. It has been made, after all, from food eaten by the pregnant cat. Now, and for the next few days, the mother needs all the energy she can find, and it will be a long time before she eats her next full meal. She cannot afford to waste anything.

New-born kittens are wet from the amniotic fluid in which they have been immersed inside the sac. After Snorkel had given

Diary · Birth

By about 10 am Barley was already born and nearly dry. I picked him up, but Snorkel snatched him out of my hand and carried him for a while as she turned round in the nest, then put him down to concentrate on expelling the next kitten – Tabitha.

Below left Tabitha has just been born and Snorkel awaits the appearance of the placenta. Below Snorkel eating the placenta. She already chewed through the umbilical cord and started licking the new kitten.

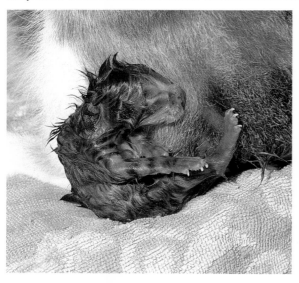

each one a thorough licking the mixture of fluid and her saliva dried much more quickly than water would from an ordinarily wet cat, and the young fur was clean, smooth and glossy.

While she is giving birth the female cat is in a highly excited state and her excitement actually helps the process. The mother's own body will control the birth of her kittens but if she is a 'maiden' – a female having her first litter – she may not know what to do next. Persians and Siamese are said to be especially prone to this kind of maternal ignorance. In any case, things may be happening fast. A cat may produce a single kitten, but litters usually consist of more than one and there may be eight or even more, and they may arrive in quick succession. Kittens can be born at the rate of anything from one a minute to one an hour, although the average is about one every half hour. There may be insufficient time to deal with one kitten before the appearance of the next, contractions may be continuing most of the time, and it may be confusing and alarming for the inexperienced mother. She may lick the first kitten but leave the rest to wander around, still attached to their placentas, and although new-born kittens may free themselves from the membranes covering their faces and start to breathe, they may still die from cold.

A new-born kitten is small. The average birth weight for a

Ginger, the fifth kitten to be born, at the moment of his birth. He is still wrapped in the membranes of his amniotic sac.

kitten is 3½ ounces (100 g). Like all very small animals its body has a very large surface area in relation to its volume. It is inside the body that heat is generated, and through the surface that it is lost. With a small body to generate heat but a large surface area through which to lose it, the kitten must be kept warm. Ignore the problem of neglected kittens and you may arrive the next morning to find tiny corpses scattered all over the place and not even inside the nest.

It is when kittens have been neglected in this way that you will need a bowl of water, because the fastest way to warm any animal is to immerse it in warm water. In the case of a kitten – or any animal with fur – you should move it around in the water to make certain the water makes contact with the skin and not just with the fur. The correct body temperature for a cat is 102°F (39°C), so that is the minimum temperature for the water and it will do no harm if it is a degree or two warmer. Of course, you must be careful to keep the nostrils clear of the water at all times. It is worth trying to revive a kitten even if it is cold to the touch and stiff, so you think it is dead. Kittens are robust, and can withstand a brief chilling.

Mothers that do not know what to do often look puzzled and dejected. If all is going well the mother should seem contented.

Even if the mother is experienced and looks after her litter

Diary · Birth

Snorkel worked hard while giving birth to each kitten and she was also very busy when each kitten had to be cleaned and dried. In between kittens she rested, looking very contented, purring and kneading with her front paws as the kittens took their first sucks of colostrum.

Lucifer, the sixth kitten to be born takes his first gasping breath. He is still attached to his placenta, which has not yet emerged.

Diary · Birth

The sixth of Snorkel's kittens was born by 4.30 pm, so that now there were five ginger males and a tabby female. Snorkel settled down with them, looking very contented, and I presumed the complete litter had been born. But when I counted them again at 10.30 pm there were two tabbies, one male and one female – seven kittens in all.

well there can be one kitten that seems neglected. It is often smaller than the rest – so it will lose body heat faster – and appears weak. It lies outside the general tangle of kittens, makes no serious effort to get closer to its mother, and even if you come to its aid and push it into a better position it still fails to find a teat. It may be injured or deformed, in which case there is little you can do to help it, but it may just be cold. A kitten that is not too seriously chilled can be warmed by resting it for an hour or so on a well-padded hotwater bottle. When the kitten is warm enough it will start scrabbling around vigorously and then you will know it is hungry. Put it back into the nest and it will fend for itself.

The most common deformity in kittens is cleft palate, which you can see by examining the inside of the mouth. If the cleft is only partial the kitten may survive, although it may have difficulty suckling. If the cleft is complete, or the kitten has some other serious deformity, it is very unlikely to live.

Snorkel was a good mother and like all good feline mothers she spent a good deal of her time licking. She licked the kittens, she licked the placentas and umbilical cords before eating them, she licked herself, and she was not satisfied with licking everything just once. Between births, as soon as she finished one round of licking she started all over again. The first kittens to arrive had all been licked many times before the last of them joined the family. Some cats spend half their time licking while they are giving birth.

She was also defensive. When Jane picked up Barley she snatched him out of her hand and carried him until the next arrival compelled her to put him down in the nest. This was the only sign of misgiving Snorkel showed, and it may have been due to her farm background. None of her previous kittens had ever been taken from her before they were at least five or six weeks old.

The first meal

The kitten who must double its weight in seven days cannot afford to waste time, and the smaller the kitten the more urgent the need. Smaller but healthy members of a litter need to catch up and so they grow faster than their bigger brothers and sisters. Human babies lose weight for a short time after they are born, but kittens do not. Feeding must begin at the first opportunity and many kittens will start hunting for a teat as soon as they are clean and dry. More usually it is half an hour or rather longer before they start, and it can be as long as two hours. It is quite common for the early arrivals to take their first meal before the last of the litter emerges.

Before it can start feeding, however, the kitten must find its mother, get to her, find its way to a teat, and stake its claim to that teat. It is not so easy as it sounds.

The young kitten must shift for itself. Until the litter is complete its mother is too busy to pay much attention to earlier arrivals, and her maternal behaviour does not develop fully until the last kitten has been born. The kitten cannot see, for its eyes are tightly closed and, although it is highly vocal, its ears are also closed, so it cannot hear. It has four limbs, but little control over them and it cannot walk. It is almost helpless, but not quite. The kitten's sense of smell is in working order and it is very sensitive to touch and especially to the warmth of its mother's body.

The kitten travels by pushing with its hind legs and making 'swimming' motions with its front legs, moving its head from side to side the whole time, until it finds its mother. Then it pushes, nuzzles and works its way along her body until it finds a teat and is able to take its first meal.

Six of her kittens have been born and Snorkel lies with them contentedly. Septimus, the last to arrive, was born some hours later.

Eating and sleeping

Diary · Four days

Snorkel left the nest only to feed and to use her tray. The kittens were not as weak and helpless as they looked. They could be quite active, and were strong enough to push siblings off a favoured teat. They moved about quite a lot within the nest, but they also slept a lot.

Like all young mammals, for the first few days of their lives kittens spend a great deal of their time eating and sleeping. At first their mother hardly ever leaves them and if she does leave she does not go far and is not away for long. Her food should be placed close to the nest, so the kittens remain in sight while she eats. Her tray should also be close to the nest – but not to her food, of course. After a few days she will start leaving them for excursions of her own, for a very short time at first but then for rather longer. If she were living wild, by this time she would need to do some hunting.

While she is away the kittens will lie all in a heap. When she returns she will awaken them, as she always wakes them, by giving each in turn a licking by way of greeting. Then they will arrange themselves all in a row, with her own body curled round them. This makes it easier for them to find a teat and it seems to quieten them.

If there are only one or two kittens they may receive more care than members of a large litter, but if the litter is small because the mother is weak, the kittens can be neglected.

While they are awake but not feeding the kittens crawl around and if they wander too far and look lost mother may have to retrieve them. They use their sense of smell to find the nest, at least over a short distance. If you take advantage of mother's absence to give the nest a good cleaning the family may become disoriented. Kittens that stray may get really lost.

Above *Septimus, now eight days old, suckling from Snorkel. He uses his front paws to steady himself but also to knead, which stimulates the flow of milk.* Right *As she is kneaded, in this case by Tabitha, Snorkel also kneads the bedding.*

Barley has been disturbed by the smell of a stranger, and hisses defensively. He is seven days old and his eyes are just beginning to open.

Feeding develops into a routine and by the second day the kittens are forming habits. Some prefer a particular part of the mother's abdomen. Others have one or even two teats they regard as their personal property. Still others are less fussy and will suckle from any free teat. The mother organizes the feeding but she cannot make all the kittens feed at the same time. They are more likely to work in sittings, first one or two, then one or two more, and by the time the last have fed the first are hungry again. Although each kitten may spend no more than one-third of its time feeding, suckling the litter may take up three-quarters of the mother's time.

Kittens do not sleep peacefully. They kick, pucker up their muzzles, twitch their noses, and their mouths may go on making sucking movements. They may also jerk their whole bodies and their legs may make small running movements. Their eyelids are sealed. If they were looser it might be possible to see the rapid-eye-movements (REM) that would suggest the kittens were dreaming (see page 51).

By the end of its first day a kitten will hiss and spit at anything strange – most usually at a strange smell or at being handled. It soon grows used to being handled, however, and can recognize a friendly human by smell. Once tamed, cats stop hissing at humans altogether.

The kittens at seven days

The kittens have developed fast. In their first week each of them has doubled in weight and they are learning rapidly how to control their bodies. Eyes are beginning to open, but at first they are pale blue and the kittens can probably see little more than movement and patches of light and shade. Probably they are starting to detect sound, although little is known about the hearing of young kittens. As yet they have no teeth, as Tabitha demonstrates.

They are learning to draw their limbs beneath them and although they cannot yet walk they can crawl by pushing themselves along, and they can hold up their heads for quite long periods. Turning corners is still difficult and they are liable to tumble, as Ferdinand has done. The dried end of his umbilical cord is still attached to him. The other kittens had lost theirs by this age. They all hate cold surfaces and strange smells and call loudly for their mother.

Barley

Septimus

Fred

See also pages 28-29

Tabitha

Ferdinand

Lucifer

Ginger

Mother care

During their first month of life, Snorkel's kittens did not leave the nest. They had no need to clamber restlessly about looking for comfort, so they never toppled out. They were warm and well fed, and Snorkel kept them scrupulously clean. They slept and fed, and slept again. And grew.

In terms of development, the first week in the life of a kitten is equal to several months in the life of a human baby, and it is during this time that the behaviour of the mother is most important. Just as babies who are neglected or ill-treated when they are very small may develop emotional or personality problems that remain with them for the rest of their lives, so kittens can be marred for life during their first few days. Kittens taken from their mothers too young, for example, may grow up to be nervous, suspicious, aggressive and less purposeful.

A kitten in distress will call for its mother with a loud, persistent wail and the mother will always respond, even though she decides on investigation that no further action is needed on her part. The wail is necessary because if a kitten is quiet the mother assumes all is well and if it has wandered from the nest she will not notice its absence.

She can recognize her own kittens by scent and will leave kittens that are not hers to their own devices. If one of her own kittens has wandered from the nest but is in no danger she may sniff at it without retrieving it. She may comfort it, however, by lying down beside it, curling her body around it, and giving it an opportunity to suckle. Then, when the kitten is quiet, she will go back to the nest and leave the kitten where it is.

She will know at once if a kitten is in real trouble because its cries will change from the usual calls for general help and reassurance to cries of distress. Usually this means the kitten is cold, and its mother will immediately bring it back to the nest.

If there is a threat to the kittens from another animal — including a strange human — the mother is capable of great ferocity. She may attack savagely and fearlessly and must be treated with respect.

Much of a kitten's exploration is random, for it cannot see properly and so unless it is heading back to the nest it will be

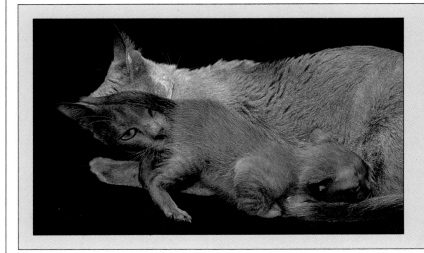

KEEPING THE NEST CLEAN

A Brown Burmese kitten being licked by its Blue mother to stimulate it to urinate, a task it cannot perform for itself. Without help it will die. This behaviour is common to all domestic cats, including those which live wild. In the wild, keeping the nest clean prevents strong odours that might attract predators.

MOVING THE FAMILY

A tortoiseshell cat carrying her one-week-old kitten. This cat grips the kitten by the scruff – the loose skin at the back of its neck – but that may be because she has done this before and knows how it should be done. Cats picking up a kitten for the first time may seize it by a shoulder or even by one leg. A kitten should not expect to be carried gently, but it always responds by going completely limp and offering no resistance. Then it tucks up its hind legs with its tail curled between them. This makes it shorter, and the mother carries it with her head held high, but it is liable to be bumped, banged and bounced all the way home. Then it will be dropped quite unceremoniously. A nervous mother may go on picking up kittens and carrying them in this way long after a more confident mother would have let them trot along beside her – as they are soon capable of doing.

going nowhere in particular. For the first few days it simply wanders in circles, but can recognize the smell of the nest if it should find it. By the end of the first week it should be able to find its way back to the nest over a distance of about a metre.

It can avoid some hazards, too, such as the edge of the nest because it is learning how to use its whiskers. It still crawls along swaying its head from side to side and if it approaches the edge of a surface its whiskers will warn it and it is more likely to turn aside than to fall.

If a wandering kitten should meet another kitten, at this very young age it will try to pass by wriggling underneath the obstruction. It is a burrower rather than a climber.

When the kittens are left by themselves they clump together in a heap and all of them go more or less to sleep. Clumping conserves warmth and provides a sense of security. Because they are burrowers rather than climbers the heap is made and

When left alone, young kittens clump. This is a heap of kittens, seven days old. Lucifer and Fred are lying on top of Septimus.

maintained from the bottom rather than the top. This means that after a little while the topmost kitten will wake because it is cold. It will slide or fall off and then burrow into the bottom where it is warmer. This disturbs the kittens at the bottom, who may object, and it also moves another kitten to the top of the heap, so that kitten wakes up cold and moves to the bottom in its turn. The kitten heap prevents any kitten from getting too cold, but it is not a particularly restful place to be.

For the first two weeks or so a kitten is unable to urinate or defecate without help. Its mother stimulates it to do so by licking it in its genital region, and she swallows the product, so preventing the nest from being soiled and also preventing the accumulation of odours strong enough to attract a predator. If you hand-rear an orphan kitten you will have to stimulate it by mimicking the mother's behaviour, perhaps by stroking with a warm, moist sponge. Unless it is able to urinate and defecate a kitten will soon die.

It is during their first week that kittens learn 'milk treading', or kneading. When they start feeding, the milk flows without prompting and as fast as they can take it. When that initial phase ends milk flow must be stimulated mechanically, and the kitten pushes at its mother's abdomen rhythmically first with one front paw and then with the other. Cats continue to find this action comforting throughout their lives, especially when they are on a warm, soft surface.

It is while they are feeding that kittens first start to purr, although purring varies considerably from one individual to another. Some cats purr so quietly they are barely audible, others very loudly, and some purr much more readily than others. The sound is made as two folds of membrane in the larynx – the false vocal cords – are relaxed and air passes across them during both inhalation and exhalation. When a kitten starts suckling and treading it also purrs. The purr attracts other kittens, and they also start suckling. The purr of the first kitten has the effect of a dinner gong, summoning the family to a meal.

Cat's milk is about ten times richer in protein than human milk and contains 70 per cent more fat, but only about two-thirds as much sugar. If you have to hand-rear a kitten do not use cow's milk, which contains more protein than human milk, but still too little for a growing cat. Unsweetened evaporated milk is much better, because the protein is concentrated in the course of processing, but if you are in doubt then it is advisable to seek veterinary advice. Kittens may be fed without too much difficulty either from a syringe or a medicine dropper.

The lactating female has to supply copious amounts of rich food and she herself needs to be fed accordingly. For the first day or two following the birth she may show no interest in food. She has eaten the placentas which provide some sustenance and she will be unwilling to leave the kittens. However, food should be offered and as soon as she starts eating she should be given extra rations in the form of extra meals rather than large helpings.

The mother needs four meals a day, which should consist of food served at room temperature – very cold food can upset her digestion.

FOSTERING

This black and white cat has two black and white kittens of her own but is fostering three tabby kittens. Females who live together and have litters at about the same time often pool the kittens, so it is not usually difficult to persuade a female to accept a kitten that is not hers. The kittens may find it more difficult to accept a foster mother, and at first they may refuse to suckle from unfamiliar teats, but in the end they take food where they can find it.

Through a kitten's eyes

Our kittens' eyes started to slit open when they were about a week old. Tabitha's were the first to open, but one or two of her brothers' did not slit until a few days later. Simultaneously, their ear-flaps, which had been folded down, began to prick.

Kittens are born blind and deaf, but with a good sense of smell and a highly developed sense of touch, augmented by their vibrissae – their 'whiskers'. At first, therefore, their knowledge of the world around them is based wholly on what they can smell and what they can feel. They learn to recognize their own mother almost at once, by her smell, and they can tell the difference between her and another adult cat. Within a few days they can find their way around the nest and can leave it to explore the surrounding area, returning safely from a distance of a metre or so.

Right from the start a cat's vibrissae are very important to it, and they remain important throughout the animal's life. A cat has these coarse, rigid hairs on the upper lip, but also above the eyes, on the chin, and on the back of the front legs. Those on the upper lip are arranged in four rows, and the cat can move the top two rows independently of the lower ones. It is these upper lip vibrissae that are the 'whiskers', and they extend sideways beyond the width of the body so the cat can use them to judge whether an opening is large enough to enter, but they are much more than measuring devices.

The vibrissae themselves are not sensitive, but being fairly rigid they act as levers. The smallest movement, especially of a long facial vibrissae, is transferred to the skin. Each vibrissa grows from a follicle, like any other hair, but one that is richly supplied with nerve endings, so the cat is immediately aware of anything that makes the tip of a vibrissa move.

At ten days old Tabitha's eyes are almost fully open, although probably she cannot yet focus them and the world appears to her as a blur. Her outer ears are starting to open and will soon be erect.

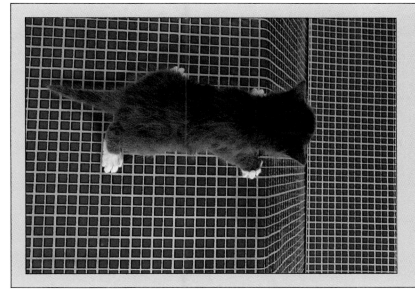

THE VISUAL CLIFF

The kitten hesitates, but the 'cliff' is an illusion. A sheet of clear glass projects past the edge of the table, covered with a checked cloth, so in spite of what its eyes tell it, the kitten can go forward on a solid surface. It soon realizes this and advances confidently, ignoring the false visual information. Having ceased to look where it is going, it falls of the edge of the glass! It would not have fallen before its eyes opened because in those days it relied on touch, not sight, and would have felt the edge of the glass in time.

Together the vibrissae add up to a very sophisticated organ of touch. They can be used to judge accurately the proximity of objects on each side of the cat's path and are brought forward to investigate what lies ahead. Vibrissae behind the front legs increase the sensitivity of the paws while the cat is manipulating an object. Nor is it only objects they can feel. They can detect the slightest movement of the air – and, of course, they work in total darkness.

The area around the tip of the nose is sensitive to small changes in temperature, useful to the adult in locating food, and to the kitten in locating its mother. The skin over the rest of a cat's body is extremely sensitive to touch – which is one reason cats find it so pleasant to be stroked – but seems insensitive to temperature. A cat can sit by a fire until its fur smoulders, but this may illustrate the insulating properties of its fur rather than insensitivity to heat.

In their second week the kittens' eyes and ears start to open and they must master two quite new senses, and quickly. If anything goes wrong so they fail to use their eyes and ears properly while they are still very young the resulting disability may be permanent.

A cat's eyes look forward, as ours do, and like us they have stereoscopic vision, enabling them to judge distances accurately. They have a wider field of vision than humans and are especially alert to small movements at the edge of that field – at the 'corner of the eye'. Unlike humans their pinnae – the visible outer parts of the ears which collect sound – can be moved, so the cat is much better than a human at judging the direction and distance of a sound.

The kittens at three weeks

When a kitten feeds, its belly swells and while it is very small its legs are not long enough to reach the floor just after it has eaten. By now, though, the kittens have grown – they weigh four times more than when they were born – and they have better control of their muscles. They can get their legs under them and are able to move around with their bellies off the ground. Their eyes are open, their ears pricked. Jane hated having to take them from the nest and put each of them by itself on a strange surface to photograph them, because they found the experience distressing and called loudly. When two met they were reassured, and even a bit playful.

Lucifer has one eye gummed shut. This is common in young kittens, whose eyes should be inspected regularly and bathed if necessary. There is a risk of conjunctivitis, which can be treated with a veterinary eye ointment but if not treated can lead to blindness or even the loss of the eye.

Septimus

Ferdinand

Lucifer

Barley

Fred

Ginger

Tabitha

Family life

The cats are a family held together by strong bonds between the mother and her kittens and among the kittens themselves. No one really knows why, but mothers often move their families to a new nest after about three weeks. They carry the kittens one at a time and after the last one has been transported, return to make sure none is left. A cat cannot count and has no idea how many kittens there should be, but she can distinguish her own kittens from others, and from other small animals.

The kittens are at about the stage of development a human baby reaches at eighteen months. They are becoming mobile, and curious about their surroundings. They feel objects with their front paws, and will soon begin hooking them towards them. Many kittens prefer using one paw to the other – they become right or left pawed. They may be starting to play among themselves.

The kittens are eating the same food as Snorkel. All the cats feed together peacefully from the same plate. Not all families are so well-mannered!

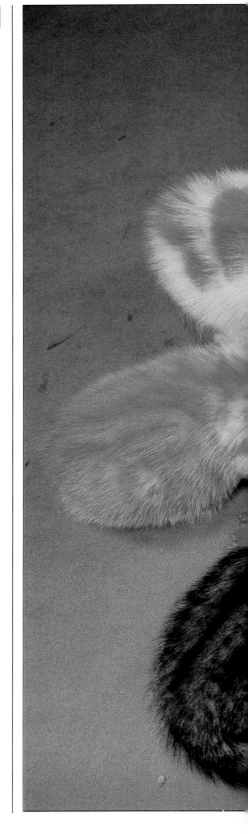

Learning to communicate

During their second month of life our kittens spent much of their time exploring and trying out their big new world, watching new movements, tasting and chewing objects, patting things with their paws, climbing on things, falling off things, pouncing, and running.

Tabitha is five weeks old, her eyes fully open and ears erect, alert as she faces something unfamiliar.

Like human babies, kittens are born with the ability to cry. The signal is effective, but limited, and they soon learn to purr as well. As they grow, also like human children, their repertoire of vocalizations increases. By the time they have mastered the full 'language' of cats they are able to communicate a wide range of messages to one another – and to humans who understand their signals.

They purr, but all purrs do not mean the same. The purr of a kitten as it starts suckling summons its litter mates. That purr may give way to a soft, smooth, more satisfied, even drowsy purring which means the animal is content. A purr can also be a demand for something that will bring pleasure. A smooth purr, but eager, insistent, and high-pitched, means the cat has seen something it wants. If it gets it, the more it enjoys itself the deeper and rougher the purr will become.

Combined with a true vocalization, a purr may also form part of a quiet, trilling greeting usually reserved for humans. Small variations in length, emphasis and pitch, sometimes with sounds added at the beginning or end, can change the greeting murmur into a pleading request, an acknowledgement or thanks, or a summons. The calling cry of a female in oestrus is a particular kind of summons.

Purrs are usually taken to indicate pleasure, but this is not always so. A cat may purr if it is in pain or if it feels ill.

Calls and cries

The murmuring sound is vocal – the vocal cords vibrate. Dr Mildred Moelk made the sole existing scientific study of cat vocalization in 1944. She found a cat uses nine consonants, five vowels, two diphthongs, and one triphthong – an 'a' that turns into a 'ou'. In addition to these there are hisses, growls, spits and screams for which our alphabet lacks letters, and there is caterwauling, which is as indescribable as it is unmistakable. The cat can arrange sounds in any way it chooses because it is not using words in the human sense, and can vary the volume, pitch and intensity.

These sounds can be grouped into 'calls' and 'cries', the ordinary 'miaow' providing the basis for calls. Each call has its particular use. It may be a demand, a protest when the demand is not satisfied quickly enough, or a kind of impassioned pleading for something the cat has seen and absolutely must have. There is also a call of bewilderment, when an excited cat is confused; a worried call; and, for the cat whose inalienable rights have been infringed, a straightforward complaint.

Cries are associated with the stronger emotions of fear and anger. A cat may make a low, dry, rather wailing cry when it is being forced into a strange and possibly dangerous situation. Screams of rage are meant as warnings, and are made to prevent fights.

Body language

When a cat – or human – needs as much visual information as possible, the pupils dilate to allow more light to enter, so a cat with dilated pupils is alert, but not necessarily alarmed. It signals alarm and aggression mainly with its ears. When relaxed the ears are erect. If the cat is frightened the ears are flattened, with the tips curled upwards. If the cat is angry the ears are also flattened, but without the curl. Twitching the ears signals anxiety.

Any sudden alarm may make a cat half crouch, ready to jump or run. This indicates fear and the more fearful the cat the closer to the ground its body will be. If it is aggressive it will hold itself high from the ground to make itself look big.

The tail, too, betrays the emotional state of the cat. A relaxed, confident cat usually walks with its tail upright. An upward flick may be used in greeting, but some cats use it as a gesture of contempt, where a human might sniff in disgust. A tail waving from side to side indicates several moods. A gentle, slow waving means only that the cat is alert – even though it may look as though it is half asleep. A quivering tail tip means the cat is excited, or anticipating excitement. A strong lashing movement is a sign of annoyance.

MAKING FACES

A six-week-old tortoiseshell kitten seeing her first dog. The dilated pupils and slightly flattened ears show the kitten feels defensive, but it is not really frightened. The pupils dilate to admit more light and so to absorb as much information as possible, and the reaction may indicate pleasure as well as alarm. Flattened ears are a defensive expression. If the ears were twisted around so more of the back of them showed, (see page 111), it would indicate aggression and the kitten might be prepared to attack. Despite their traditional enmity, dogs and cats often tolerate or even enjoy one another's company and a cat is seldom in real danger even if a dog threatens it. It can outrun a dog over a short distance and can climb or jump to safety.

The kittens at six weeks

As their faces and postures show, the kittens are now alert and interested in the world around them and in one another. They play with soft toys, such as balls of wool, and are no less fascinated by leaves, feathers, and the twitching tails of other kittens. They can walk and run well, although they still tend to tumble when cornering fast and, as Tabitha demonstrates, they can groom themselves. They are distinct individuals with personalities and make their own decisions on important matters such as when to eat. They demand food rather than being called to it.

The kittens are very used to being handled and are no longer distressed by being put on a black surface to be photographed.

Barley's yawn displays his milk teeth. These appear when the kitten is two or three weeks old, the incisors first and then the canines. Kittens have no molars. These appear at about six months, along with the rest of the permanent teeth.

Ginger

Fred

Ferdinand

Lucifer

Barley

Septimus

Tabitha

Snorkel's kittens have always had nice table manners. There was hardly any jostling and meanness around the dish. Even now they are getting quite big, they eat amicably from the same dish.

The plate is empty, the others have left, but Ginger, Septimus and Lucifer remain, ever hopeful. They always seem to be hungry.

Weaning

From the time they are about four weeks old kittens start taking an interest in the food their mother eats. They are playing well by then, are curious about everything, and they have teeth. Weaning can commence, but the interest in solid food comes from the kittens themselves. Their mother does nothing to persuade them.

The mother, an adult house cat eating food that has been prepared for her by humans, need take no special precautions over what her kittens eat. Her food will not harm them, and in any case the desire to eat it is theirs and all she has to do is let them eat from her dish. The kittens learn from their mother, in the sense that it is the sight of her eating that encourages them to imitate her and eat in the same way, but her role is passive. They will take only a very little food at first, experimentally, and then more at each meal. It is even more important that the food presented to the family be at or slightly above room temperature. Very cold food will upset small stomachs.

If the family were living wild there would be more for the mother to do. She would be hunting or scavenging for her food and the kittens would have to be taught to distinguish what is good to eat from what is not.

She would begin by bringing prey back to the nest. This sounds little enough, but a hungry cat has a strong urge to eat

TROUGH COMPETITION

Being a member of a family is a matter of give and take. Among kittens the emphasis is more on taking than giving. This is called 'trough competition'.

Kittens are always ravenously hungry and like some children they will eat until no food is left, even if it makes them ill, and the kitten who gets to the bowl first

gets most in their game of 'winner scoffs all'. You can fend off the competition with a paw and a growl, or beat it by being first and eating fastest.

food when and where she finds it. If it might be dangerous to remain in the open she may carry the food to a safe place, but her urge is still to eat it herself, and she will drive away any other cat that tries to take it from her. When she has kittens she overcomes those feelings, carries her food to the nest, and presents it to her kittens.

She drops it where they can see it and gives an urgent cry, quite different from any other, which means something like: 'Quick! Look what I've got!' She will let the kittens examine her prey, sniff at it, and then she will eat it in front of them. In this way they learn that small things looking and smelling like this can be eaten – by mother, at any rate. After it has happened a few times with different prey, the mother introduces a new version of the summoning cry, so that now the kittens understand two messages, and the meaning of the first one has changed. The original cry now means: 'Here is something small, harmless, and edible.' The second means: 'Here is something big, dangerous, but edible if you get it before it gets you.'

As they gain confidence, the kittens will pat at the prey, or even jump on it, and starting with the braver ones, they will taste it. The mother may throw the prey around, toss it in the air, and generally abuse it – perhaps to demonstrate how prey may be dealt with – before allowing the kittens to have it. The whole operation serves to familiarize the young with the kind of food that must sustain them for the rest of their lives.

The second stage, after this lesson in recognizing prey, is to leave the food for the kittens to eat, and the behaviour becomes more complex. Now there are three kinds of prey. There are the first two, which the kittens may eat, and then there is a third kind, the prey that mother eats herself and that the kittens cannot have.

Mice on the kitchen floor

So far as anyone can tell the adult cat is stimulated to behave in this way by the kittens themselves and not by hormonal changes in her own body. Females who have no kittens of their own will bring food to another cat's kittens in this way. It has been known for a cat with kittens to adopt a litter of older kittens belonging to another cat and to bring food to both, but at a different time for each litter. In a still more remarkable observation, among a group of related farm cats whose behaviour was studied closely over a year, one female was seen to bring food and present it to her sister who was raising a litter. Deliberate sharing of food between adults is an extension of the supplying of food to kittens. It is probably uncommon, but it can happen.

It is much more common for a female house cat, especially one that has been neutered, to bring prey and present it to her human companions. She will bring in whatever she has caught, usually mice, and lay them out for inspection. When she does this she is not seeking praise or admiration. She is not trying to show off her prowess as a hunter, and although she will enjoy being petted because she always enjoys it, that is not why she brought the prey home. She wants the humans to examine the prey, to demonstrate that they are taking an interest, because she is teaching them. She has no kittens so she invents substitutes. The humans become her surrogate 'kittens' and, good mother that she is, they must be taught to recognize food when they see it. As a cat owner, you should enthuse over the prey to show you have learned the lesson. You will not be expected actually to eat it and with any luck the cat will forget to follow up the first lesson with the second.

During this first lesson all the prey has been brought to the nest dead. When the kittens are very young it would be far too risky to expose them to live prey, for even a mouse can give a kitten a painful bite. The next stage in their education is to present them with live prey, and that is the beginning of their training as hunters.

Diary · Seven weeks

Snorkel's family all fed from the same dish. In some families the first to arrive hogs the lot while the rest crouch around watching, and stay hungry (see page 37). The only solution then is for each kitten to have its own dish.

Litter training

When the kittens start eating solid food the composition of their body wastes changes and their mother stops cleaning them. They can control their bodies by now and must find somewhere to relieve themselves. Some find it more confusing than others and may leave puddles and droppings anywhere, but most take readily to using a litter tray.

While the kittens are with their mother the obvious tray to use is hers, which will have her smell, and perhaps that is how they learn. Hand-reared kittens also learn to use a tray, however, from as young as three weeks, and they can have had no maternal example to follow. Farm kittens that are moved into a house will also use a tray from the first time they are placed in one, although they have been used to the soft earth. There are substances you can buy in pet shops to sprinkle on fresh cat litter to encourage young kittens to use the tray, but in most cases this is unnecessary. You should buy a large tray because even a small kitten will grow, and if the tray is too small you may have to deal with frequent near misses.

The mother does nothing to retrain her kittens. They train themselves, but they must have somewhere they can dig. Many animals bury their droppings, but unlike some which scratch enthusiastically even if the surface is hard, cats dig only if digging is feasible, although they may do a bit of ritual scratching. Cats that are used to being outdoors will pick certain places and use them regularly. Indoors they will use a tray, but if there is no tray they will relieve themselves anywhere. Once that behaviour has been allowed to establish itself it may be impossible to retrain them.

Diary · Seven weeks

All the kittens loved the feel of the litter and scraped and dug and played in it even if they then occasionally climbed out of the tray and puddled elsewhere. The kittens also ate the litter, crunching it up like biscuits. I let them eat the litter but if I thought it was doing a particular kitten any harm I would talk to a vet about it.

PLAY IN THE TRAY
Litter trays have many uses and kittens are endlessly inventive in their search for new toys. This kitten, seven weeks old, is using the tray as cover from which to stalk his brother, who is using the door for the same purpose. It is best to place the tray on a surface that is easy to sweep. Dry litter can get trodden everywhere and when the tray is used for romping and chasing games litter will be spilled.

Family greetings

Snorkel was still quite happy to let the kittens suckle, although they were eating solids and could well be weaned.

When two humans meet they greet one another in a very ritualized way, with certain gestures and expressions that both of them recognize as signals of friendliness. Cats, too, greet one another when they meet, but among members of a family the greeting has an additional purpose. They mark one another with scent, so that while every individual smells different, members of a family smell enough like each other to confirm their identity. By the time they are six or seven weeks old kittens must be able to greet each other, and their elders, politely. If they are house cats they soon learn to greet humans in the same way.

A cat greeting another cat must approach it head-on. Attacks usually come from behind, so the cat that faces another means it no harm. Then it must approach closer until the two can touch noses, a gesture which means 'Hello, do I know you?' The sniffing may extend to anal sniffing if there is any doubt.

This may be difficult for a kitten approaching an adult, or impossible for any cat approaching a much larger animal. The alternative is to flick the tail upright, make a vocalized purring murmur, and sometimes to give a little, stiff-legged hop by raising the front of the body and lowering it again.

Snorkel comes in and is greeted by her kittens with friendly tails-up.

The greeting having started it can now proceed in two ways, according to the affection the cats wish to show one another – or the cat wishes to show to another animal. For the more casual greeting one cat walks past the other so their flanks barely touch. As they pass both cats hold their tails almost upright, but with a slight curve in the direction of the other cat. The more affectionate greeting involves one cat rubbing its cheek hard against the flank of the other, and this may develop into mutual greeting in which the cats rub their faces against each other. This is the way a cat often greets a human it likes.

The greeting is a sign of affection, but it is by cheek rubbing that the cats exchange scent. A cat has scent glands on the lips and chin, on the sides of the forehead, and along the tail. Humans cannot smell the scent, but if a cat jumps on you, touches you nose-to-nose, and then turns its back and raises its tail, you are expected to sniff at the base of the tail.

Left *Snorkel is licking Lucifer's fur as a mark of affection, not because he needs washing, while Lucifer starts to suckle – despite Ginger playing with his tail.* Below *Although the kittens are more than seven weeks old, Snorkel still allows them to suckle.*

41

Kittens at play

Some mothers play with their kittens, others remain aloof. Snorkel was one of the aloof kind, though no worse a mother for that. Some kittens receive visits from their fathers, who also play with them, but other fathers will have nothing to do with their offspring. Older brothers and sisters often play with the kittens, and spend some of their time 'kitten-sitting' while mother is out, but Snorkel's litter had to play by themselves.

Play begins early. At first kittens are either hungry or tired, and when they are alone they sleep in a heap. This clumping behaviour continues but after a couple of weeks or so the clump itself changes and the ideal position for a kitten is to sleep with its chin resting on its neighbour. Attaining this is difficult and involves much wriggling, but when everybody is happy, the whole clump falls asleep together. However, as has been mentioned on page 24, they wriggle in their sleep, kick one

Lucifer and Tabitha play-fighting, at eight weeks old. No matter how fierce it looks, kittens never hurt one another while playing.

another, and are liable to wake up at the bottom, worm themselves out, and climb back on top. What started as a peaceful rest often degenerates into a scrap.

The kittens are learning to control their muscles, and as the muscles grow they are flexing them. There is no one to teach a kitten to stand or walk or run, but it must learn, and one of the things that is learned early is the vertical jump. Stretch all four legs suddenly and together and you fly up in the air. That is jumping, but it is more fun if you can jump somewhere, and that somewhere tends to be another kitten. A kitten will resist another kitten which lands on top of it, and so jumping leads to fighting.

The victim of this sort of attack usually rolls over on to its back. That is the best position for fighting because it frees all four paws and the jaws. The forepaws are used to hold the opponent while the hind paws rake it, and bites may be added for good measure. Kittens never hurt one another. When they are very small their claws cannot be retracted, but they are tiny, and so are their teeth, and incapable of inflicting damage on an animal protected by fur. Later, when the young cats could injure one another, they restrain themselves from doing so. Their rough-and-tumble fights look fierce, but they do no harm.

When its mother washes a kitten she rolls it over on its back to deal with its underside, and she also licks it to stimulate it to excrete. These are pleasant sensations and cats never lose their love of having their undersides stroked while they lie on their backs. It is because this pleasant position is also the best fighting position that a cat whose tummy is being stroked may decide at any moment to turn the encounter into a play-fight.

Stalks and pounces

Jumping at any kitten that comes within range is all very well, but it soon palls and the kittens develop more sophisticated methods of attack. They will stalk one another. The stalking leads to a final chase, but the kittens have learned they must not hurt one another, so the final attack is usually abandoned. Such an abrupt halt makes a kitten fall over and by the time it has acquired enough control to avoid falling it is learning to use cover, to crawl on its belly, and to poise its body, tail waving purposefully, for a final leap it may or may not make.

Stalking and pouncing is play that mimics hunting and dealing with prey, but play-fighting continues. If a cat is threatened by an opponent larger and more powerful than itself it will try to escape, but escape may be impossible. In that case there are two alternatives. If both combatants are cats the weaker may adopt a submissive posture, crouching as low as it can to the ground and exposing the back of its neck – its most vulnerable point. This is usually enough to defuse the situation. The stronger cat accepts the 'apology' and allows the weaker one

Diary · Eight weeks

Snorkel was a very tolerant mother and appeared very fond of her kittens, but did not play with them. There were probably too many of them, and they had playmates enough without her. When they had all woken up after their feed, she would simply get up and either jump up where they could not reach her or go outside where they had not yet learned to follow.

to withdraw. Kittens must learn this – they are very small and weak, after all – but it takes a little learning and it is not very exciting. They prefer the other option, which is to make a surprise attack in an attempt to catch the opponent off-guard.

If this is to have any chance of success the attacking cat must make itself look as big and alarming as it can. It stands on tip-toe, arches its back to give it a little more height, and raises all its fur. In all mammals, each hair follicle has muscles that allow the hair to be pulled into an erect position. With all its fur standing on end and its tail looking like a bottle-brush the cat is considerably larger (see page 47). At the same time, because the fur is not solid and all the hairs are not precisely the same length, the outline of the animal is blurred. This, too, can confuse an opponent used to seeing a sharp outline at which to aim. The confusion is likely to last only a fraction of a second, but it may be enough.

Ears laid back, snarling or even howling, front claws extended, the kitten is now ready to launch its surprise attack. It must not lose the advantage of apparent size, which it would do were it to advance the usual way, head-on, so it advances sideways, the hindquarters seeming to move faster than the forequarters, in a kind of dancing movement, with its weight thrown back so it can raise its front paws. The attack is not fast – if the dangerous opponent withdraws the attacker will be more than satisfied since its object is to escape – but it is truly frightening.

1 *Barley playing with a realistic-looking toy mouse. He has 'caught' it, carrying it with his head held high. This stance keeps the prey well clear of the ground.*

2 *The mouse is held by the back of the neck. This immobilizes it and, in particular, it keeps its jaws away from the kitten.*

The excited kittens will launch such attacks, two of them dancing towards one another. Sometimes they miss and each kitten dances right past its opponent. Sometimes they do not miss. They bump into one another and fall over.

Playing with toys

Stalking, hunting, pouncing, attack and self-defence are very important, but they are not all. The adult cat will have to catch prey and so it must learn how to manipulate objects. For that the kitten needs toys, and all kittens play with toys. The toys must be small and light, so they move easily, and the best ones are soft, so the kitten can grasp them with its claws. The traditional ball of wool may be the best toy of all, and a kitten will play happily with a toy pulled or swung along on a string by a human.

The kitten does not share its toys, although they may be stolen by a rival. While the toy is on the floor – ideally on a smooth floor rather than a carpet – the game consists in stalking and pouncing, sometimes biting, and then batting the toy away with a paw and chasing it. If it disappears under the furniture the kitten will try to hook it out with a paw. If you throw it along the floor the kitten will chase it. This is all rehearsal of the movements it will make when it pursues a mouse.

If you hang the toy so it swings free, or dangle it, the game changes. It will give chase and jump to catch the toy, trying to

3 *Alert and cautious, Barley taps the mouse to see what it will do. If it moves it is still alive and he will catch it. If not, it is dead and he will 'eat' it.*

4 *It did move, and for the moment it has escaped. Now Barley has to prepare a new attack and launch it quickly, or the mouse may be gone for good.*

seize it between its forepaws. This game has nothing to do with mice, but it has a great deal to do with the small prey animal that runs along the ground like a mouse, and then leaps into the air. Thus the cat has to attack not at where the prey is, but at where it anticipates it will be. It must jump, grab the prey in its paws, and bring it to the ground. This is the way a cat catches a bird.

While playing with a toy that is not attached to anything, a kitten will often grab it with one paw and throw it up in the air. After some practice the throw is usually over the kitten's head towards the rear, and the kitten will follow the flight of the toy and pounce on it when it lands. The kitten is trying its skill at fishing.

Fish do not figure large in the diet of most cats that fend for themselves, but no opportunity for a meal can be overlooked. The fishing cat sits by the water's edge and waits for a fish to approach. Then it will dip in a paw, with the claws extended, catching the fish from below. It will then flip it out on to the bank.

Why do adults cats sometimes 'toy' with live prey? There is no real explanation, but the behaviour is infantile. The cat is behaving with its prey as it used to behave with toys when it was a kitten. It is by retaining some of their infantile characteristics that cats are able to adapt themselves to living with humans and being handled by them. Perhaps 'toying' is part of the package, a dark side to the 'kittenish' behaviour we like.

5 *A failed pounce! Barley made his leap and has landed, claws out, ready to hold his prey, but either the mouse was too quick or Barley missed.*

6 *He has it now, held by the neck, looking somewhat the worse for wear, and glares fiercely at the camera. His pupils are dilated – a sure sign of alertness.*

THE 'WITCH'S CAT'

This eight-week-old Blue Abyssinian kitten is playing at fighting in a tight corner. It makes itself look as large as possible by arching its back and erecting all its fur. Its tail looks like a bottle-brush. The erect fur also helps to obscure the outline of its body. Its ears are back, its head forward, and it presents itself side-on. Its advance will be a kind of dance, looking as though its back legs are advancing faster than its front legs, and its front paws, claws extended, will be ready for a slashing attack.

7 Lest he be robbed, Barley turns away, but slowly and cautiously. From the corner of his eye he can detect the smallest movement.

8 Now he is nervous and has decided he had better take his game somewhere else. His crouching stance, with his body held low, shows he is unsure of himself.

The kittens at ten weeks

The kittens are now much more like young cats. They are adventurous and have lost their dislike of the strange black surface on which Jane photographs them. Indeed, Barley so enjoyed the feel of the black velvet he had to be prevented from tearing it to shreds.

Their senses are fully developed and they have complete control of their muscles. They were able to balance on their haunches from about five weeks and now use this position to reach upward, as Ginger is doing to grab at

Barley's tail. Any object will be examined in case it might be worth chasing or fighting, but the ball of wool is the perennial favourite – and the wool will break before an entangled Ferdinand can be injured. Fred exhibits all the grace of a cat on the move.

Lucifer is playing with counters. He can see the colours, but colour is unimportant to any cat. Cats are mainly nocturnal, and react to sounds, smells, and small movements, not to colour which not even they can see at night.

Septimus

Barley

Ginger

Tabitha

Fred

Ferdinand

Lucifer

Lucifer wakes up and starts a huge yawn to draw in a charge of oxygen to supply energy to his relaxed muscles.

Diary · Eleven weeks

At nearly three months old and now getting large the kittens' sleeping habits were still much the same as when they were little. Not that they clumped together now but they often slept clustered in quite a heap, paws around each others' necks.

Sleeping and waking

Cats are renowned for their ability to sleep as and when opportunity offers, for hours on end, or for periods so brief they have given our language the word 'catnap'. This love of sleep is not peculiar to the domestic cat. The lions, tigers and other large cats are also highly proficient in the art.

We neurotic humans may think that time spent sleeping is time wasted and that the cat is an idle animal, but unlike us it has no compulsion to fill every moment with physical or mental activity. It does have a need to conserve energy, however. Energy is provided by food. The cat is a carnivore, and meat is the richest kind of food, but it is more difficult to obtain than plant food. The cat needs to eat less, and so to spend less time eating, than a herbivore subsisting on a less nutritious diet, but it can never be entirely certain about its next meal. There are long periods, therefore, when it is not seeking food and is trying to make the food it has last. There is nothing for it to do, so it does nothing. Scientists who have observed cat behaviour closely have concluded that their subjects spend a great deal of their time doing nothing in particular.

Cats sleep from boredom, and they may also use sleep to escape from unpleasant situations. Cats that have spent all their lives in cages may tolerate such confinement, but a cat finding itself in a cage for the first time may sleep much more than usual, as a way of hiding. Well fed, contented house cats sleep more as they grow older.

Dreaming and dozing

Most animals have three distinct types of sleep, a shallow, light sleep, a really deep sleep, and that half-sleeping half-waking state from which we may slide gently into sleep or wakefulness. Cats spend happily relaxed hours in this state.

When they are very young, kittens have only two mental states. They are either awake or asleep, and the only sleep they know is deep sleep. The body temperature rises, the heart beat slows, the blood pressure falls, and the muscles relax. This is the kind of sleep psychologists call 'paradoxical,' or 'REM' sleep, REM standing for 'rapid-eye-movement.' The sleep is 'paradoxical' because although the sleeper is more deeply unconscious, its brain-wave activity is greater than during light sleep and probably the brain is as active as it is when the sleeper is awake. Brain-waves, in humans as well as in cats, are measured simply and painlessly from electrodes fastened to the head by wax – the process involves no surgery. During paradoxical sleep the cat's eyes are just slightly open, its pupils are widely dilated, and the third eyelid, the 'nictitating membrane' is withdrawn further than when the animal is awake.

Just before the muscles relax, the eyes may begin to move. They move slowly, but this is called 'REM,' and the ear muscles are active as well as those of the eyes. The sleeper may purr,

Lucifer waking. Cats may wake up briefly after dreaming to yawn, stretch, then go back to sleep, but this is a full, waking-up stretch.

It begins by arching the back, then extending the front legs. After that Lucifer turns around, walks a few paces, and stretches his back legs. This behaviour is fairly stereotyped it is nearly always the same.

growl or murmur, limbs may twitch, and a sleep-walking cat may stalk invisible prey. The cat is dreaming. All animals dream, and if they are prevented from dreaming they dream more as soon as they can, to compensate. An adult cat spends about one-third of its sleeping time in paradoxical sleep.

A young cat starts sleeping lightly when its nervous system reaches a certain stage of development, at one to four weeks old. In this state its temperature falls and its brain activity is reduced, but it is less relaxed than in deep sleep, and its hearing actually becomes more acute.

The father's return

Diary · Eleven weeks

When the kittens were eleven weeks old Snorkel seemed to start calling again. This marked a turning point in the kittens' lives. Instead of returning their friendly greetings and letting them suckle as she always had before Snorkel turned on them fiercely. Fergus reappeared among his family, again lured by Snorkel's calls and scent. At first, the kittens were afraid of him and he was inclined to spit at them, but gradually they got to know one another and became good friends and even playmates. Snorkel was pleased to see Fergus but almost immediately rebuffed him. She was in false oestrus and would not tolerate close approach. Unfortunately she died from a tumorous kidney several weeks later.

Instead of the usual affectionate maternal greeting, Tabatha gets a fierce rebuff from her mother. Weaning is now complete.

Weaning may begin almost accidentally as kittens join their mother while she eats, to play or suckle perhaps, and chance to taste her food. If a human puts down a saucer of milk the kittens may examine it out of curiosity. There will be some choking and sneezing at first, until they learn how deeply to push their muzzles into the liquid. As they take more solid food they continue to suckle, but they are growing, and so are their teeth, and sucking is soon accompanied by biting. Probably it is pain that makes the mother decide enough is enough, or she has a false oestrus, and when that happens she will refuse to allow any more suckling.

This marks the end of infancy and relationships change. The kittens are still young and have much to learn. The family stays together as a coherent social group but if you were to compare it to a human family you might say the children are now ready for school. They can wash and feed themselves and tie their own shoelaces, as it were.

Life with father

It is the stimulus of sucking that causes the mother's milk to flow and when that flow ceases hormonal changes in her body begin to prepare her for another litter. If she is a well-fed house cat she may come into oestrus after about three weeks. If the family lives wild it will be longer, and the family will have about a year together before the next kittens are born.

It is around this time that father may pay a visit. Some males visit their kittens regularly and play with them, but most take no interest whatever in their perhaps several families.

A male cat hunts within a range that is much larger than the range of a female and includes several female ranges. The female ranges may be crowded and often overlap, so within his exclusive preserve the male has a number of females to visit. When he calls he is greeted, there may be displays of affection such as some mutual grooming, but if the female is not in oestrus the male will not stay. If she is in oestrus, of course he will mate with her. He will probably ignore any kittens unless they annoy him by trying to play with him when he is tired or has better things to do, and then he will growl at them or bat them away with a paw. To most kittens, father is not very important.

The mother's first oestrus after weaning may be preceded, about a week earlier, by a false oestrus. She will call, roll, present herself, and the male will be aroused, but if he tries to mount she will turn on him, snarling and clawing. After a few rejections he may try to mount a kitten. The attempted mating fails, but can be dangerous. While mating the male grips the back of the female's neck, quite gently, with his teeth (see page 115). He will grip a kitten in the same way and if the kitten struggles his grip will tighten, sometimes with fatal consequences. The kitten usually keeps still and escapes unharmed.

Fergus and Snorkel meet again for the first time since the birth of the kittens. They seemed delighted to see each other. Fergus rubbed his head against Snorkel's chest and she licked the top of his head, left, *but suddenly she lashed out,* below, *warning him to keep his distance.*

53

Tabitha's progress

Infancy is a dangerous, difficult time that must be passed through as quickly as possible and, sadly perhaps, kittens do not remain kittens for long.

At one day old (1) Tabitha is blind, deaf, and only 4 inches (10 cm) long. At the end of a week (2) she is 5 inches (12.5 cm) long. Her eyes and ears have started to open and she is gaining some control of her limbs. After three weeks (3) she is 6.½ inches (16.5 cm) long. She can see and hear, and her legs are held beneath her, so she can walk properly. At six

weeks (4) old she is 8 inches (20 cm) long and a playful kitten, able to run and jump, playing with toys and romping with her brothers. At ten weeks (5) childhood is over, and the kitten is a young cat, 10 inches (25 cm) long, eating solid food and venturing into the world outside.

5

Exploring and learning

All young cats are highly inquisitive. They investigate everything, explore everywhere, and sometimes their curiosity leads them into scrapes. Such an appetite for discovery marks the cat as highly intelligent, but animal flattery apart its curiosity has a very practical purpose. In a world it is experiencing for the first time, where everything is new, it has half a year, perhaps less, in which to prepare itself for adulthood by learning how to find food and keep out of trouble. It must develop its hunting skills by practice and experiment, and it must familiarize itself thoroughly with its surroundings.

The inquisitive cat is looking for, and remembering, sights, sounds and smells that mean danger, the kind of places in which danger may lurk, and places where meals may be found. The playful cat is rehearsing the actions by which it will catch those meals.

Barley is exploring the wide world outdoors. He looks as though he is playing, but this is part of his education. The twig has an interesting smell, perhaps left by a dog. Is danger nearby?

Learning the cat business

The kittens were ready to be allowed outside. They were now fully innoculated as well as robust enough to extricate themselves from most difficulties. One of the first of these that Septimus encountered was to do with the pond. Not wanting to get off the narrow parapet in the middle by going back the way he had come, he stepped on to some floating weed supposing he could walk across the water. After flicking his wet forepaw a good deal he next tried to jump across with even wetter results.

During the weeks a kitten spends in and around the nest, playing with toys, play-fighting, running, stalking, pouncing and laying ambushes it is receiving its equivalent of primary education. By the time it is about three months old its senses are keen and fully developed, and its strong, agile body obeys its owner's commands. It is time to move on, to start applying new skills in the real world, to prepare for adult life as a hunter, and there is much still to learn.

The outside world is strange and the cat does not see it as a human sees it. The cat is small and for much of the time its view is obscured by plants taller than itself. If it moves into open ground, where it can see further, it can also be seen, and risks being attacked. It prefers not to move far from protective cover and although its eyesight is sharp, it explores less by vision than

Above *Septimus tries to jump across the pond but miscalculates and lands in the water.* Left *Bedraggled and disgusted he hauls himself ashore.* Right *Cats learn quickly and remember what they learn. The next time Septimus wants to cross the pond he leaps, and stays dry.*

by its senses of smell, touch and hearing. It is by scent that it knows another animal has passed, and the freshness of the scent tells it how recently. It feels the least breath of air and hears the slightest movement. The hairs in its ears are so fine scientists believe they may be subject to the same quantum forces that affect subatomic particles, causing them to make incessant random movements which the cat can hear. These drown out quieter sound and so mark the threshold of its hearing. There are eddies in the blood flowing through the veins and arteries close to its ears and the cat may also hear those.

As it explores, the cat memorizes its surroundings, building up a mental map of sounds, movements, smells and, at close range, sights. The cat discovers places where it can rest, and even doze, hidden by surrounding vegetation, protected from

Dogs and cats often get along together perfectly well. Tabitha, three months old, greets Honey with a friendly upward flick of her tail and nose-touch.

the wind, but exposed to the warm sunshine. The cat notes the paths worn by mice and voles as they scuttle to and from their nests on foraging expeditions. The cat remembers the trees which it can climb easily, and where hedges and fences that have gaps it can squeeze through, but that will check a larger pursuer.

It examines carefully but cautiously any object it meets for the first time, sniffing, touching, patting, as it places the object in one of the two categories that matter – food or hazard. It is very suspicious of any animal that has neither fur nor feathers, probably because such animals usually have no clearly visible neck, which is where a cat aims its attack on prey but also because any animal might be dangerous. Snakelike animals are treated with respect by most mammals.

Cats cannot lose body heat by sweating, as we do, but they can pant. On a very hot day Septimus finds a cool, shady spot in a pile of logs, and stays there awhile, panting.

This is learning by trial and error, but cats also learn by watching others and they can learn very quickly and remember what they have learned. Some are quicker and more intelligent than others, of course, but a really smart domestic cat may be as good at solving problems as a monkey or even an ape if the solution gives it something it wants.

Exploration and learning by trial and error are essential. When the kitten is old enough to start serious expeditions away from home it is also time for it to be vaccinated. A young kitten is protected from disease by the natural antibodies in its blood, but this passive immunity passed to it from its mother has waned by about three months.

The three most common diseases of cats are feline enteritis and two upper respiratory viral infections known in Britain as 'cat flu', and all of them can be prevented by vaccination. One of these, feline viral rhinotracheitis (FVR), produces symptoms much like 'flu in a human. The other, caused by feline calicivirus (FCV), produces a cough and ulcers around the nose and mouth. Both may lead to secondary infections and complications. Left untreated, most healthy adult cats will recover, but they may remain symptomless carriers of infection and so dangerous to other cats – including their own offspring. The diseases can also kill indirectly. A cat relies on smell and taste to tell it what is wholesome. Like a human with a cold a cat with either form of 'flu' cannot smell or taste things, and this loss, combined with general distress, may make it refuse to eat, so it starves to death.

Diary · Three months

In the heat wave of the summer I noticed that the kittens were unhappy out in the sun. Cats seem to suffer much more than dogs in these conditions.

61

The agile climber

Cats are born climbers, but are very much better at going up trees than they are at coming down. Some of our kittens were better at descending than others. Poor Lucifer was always the one stuck at the top of the ladder or the apple tree, wailing that he could not get down, whereas Ginger was the ace climber who soon learned all there was to know about trees.

Cats like being in high places. This is partly a matter of social prestige. Just as a cat signals its submission to a superior by crouching low to the ground, so it indicates its superior status by raising itself as high as it can. It likes to look down on its inferiors and when the house cat perches on a shelf or on top of a high cupboard to survey the room and its human occupants, that is partly what it is doing. It regards the humans as inferior.

All cats can climb, more or less. Even a lion will struggle into a low tree now and then, if there is one with a branch capable of supporting its weight, but it is the smaller, lighter cats that are the real experts. Leopards will take their prey into a tree, where they can store food and eat without fear of interruption, but the most prodigious climber of all is the ocelot, a South American cat about the size of a spaniel dog.

The ordinary cat climbs well enough, but the ocelot is much more at home above the ground than any domestic cat. It ascends a tree trunk by spiralling around it. It can drop from branch to branch, hang by its back feet while it examines – or even plays with – an object with its front feet, and then swing itself up again. When it is time to come down it descends a tree headfirst. The house cat may be a good climber, but it is not that good. Yet the ocelot and the domestic cat are very similar. They have bodies built to an almost identical pattern. Why is the ocelot so much more skilful?

Fingers and claws

The difference between a house cat and an ocelot lies in their muscles, ligaments and joints. There is no essential difference in their skeletons. The ocelot can rotate its hind feet so the toes point to the rear, it can twist its ankles so its hind feet can face inward or outward, and it can close its hind toes so they grip a narrow branch or vine. The house cat can do none of these.

Apart from cheetahs, all cats have front claws that are retracted when relaxed. When extended – or protracted – they are sharper, longer, but not so strong as the hind claws. When it climbs the cat can support its weight with the hind claws while it grips and pulls upward with the front claws. Ascending a vertical tree trunk is easy, even for the house cat that goes straight up rather than spiralling. Descending is a different matter. The ocelot can rotate its hind feet, so the curved claws can grip the trunk and support its weight, then hold on with its front claws just long enough to bring its hind legs forward to a new hold. This allows it to descend headfirst.

The house cat cannot rotate its hind feet and if it tried to hold on with only its front claws it would overbalance, somersault, and fall. Provided the ground surface below was fairly even the fall would not harm it, but the cat would not know that, and cats are frightened of falling. They are expert at judging distances and from high in a tree the ground looks terrifyingly remote.

Opposite Septimus climbing among poles, demonstrating just how flexible a cat's body is.

Below *Ginger has no difficulty in climbing up the trunk of the apple tree or,* right, *in moving about among the branches. Provided he can support his weight with his hind feet and grip with his hind claws, he feels safe.* Far right, top *Coming down is much more difficult. He has to descend tail-first in a series of jerky little movements as he clings with his front claws while he slides his back claws down to feel for a hold, glancing anxiously over his shoulder at the ground.* Far right, bottom *The descent ends with a jump, but he has to turn, still on the trunk, and push with his hind feet before he falls out of control.*

The only alternative is to descend backwards, tail-first, using all four sets of claws to grip. This is only slightly less alarming for the cat, because it cannot see where it is going. It has to feel for its toeholds and make do with glances over its shoulder to measure its progress.

Stairs, steps and ladders can be negotiated more easily, because the cat can get enough of its body between the treads or rungs to support its weight with its hind feet. However, the descent of a pole or tree may well often be a slow, clumsy, undignified affair.

Lucifer is stuck on top of the steps but Fred has learned to support his weight on the treads and can descend headfirst. Compare this picture with the action photograph on page 95.

It may be possible for a cat to make its way down a tree by jumping from branch to branch. Cats jump well, but only if they have a firm surface against which they can push. Once it leaves the thick main branches of a tree there may be no such surface, and jumping may be even more hazardous than climbing.

Some cats learn to descend more quickly than others and some are better at it than others, perhaps because their hind feet are just a little more flexible. This means they can turn them enough to get a more secure grip, and as with anything else performance improves with practice. Not all cats are brave enough to do much climbing except when they are chased into a tree.

The cat stuck on a roof or in the top of a tree, wailing miserably with fear, is a familiar sight that accounts for many of the calls received by fire brigades and animal welfare organizations, but in the vast majority of cases the cat will find its own way down eventually. Its hunger will overcome its fear and it will suffer no harm.

Flexibility and balance

A cat may have trouble descending a tree, but it will walk confidently along the narrowest branch or the top of a wall. Its sense of balance is renowned.

Before they were used for hearing, the ancestors of our ears were organs of balance, made, as they are still, of tiny bones whose movement we detect. Cats have an excellent sense of balance, based like ours in their ears, and for reasons no one has discovered they seem not to suffer from motion sickness when they travel. It is one thing to sense when you are off-balance, however, and quite another to prevent yourself falling.

The cat uses its tail to help it balance, like the pole carried by a tightrope walker, but that is only part of its secret. In any case the tail is of limited use, except when the cat has to change direction quickly and needs to throw its weight to one side. The cat relies much more on its skeleton.

It has thirty vertebrae – five more than a human – and the muscles of its back are so strong that the spine can be extended or contracted much more than that of a human. It can be twisted through a half circle, so one half faces in the opposite direction to the other. Its shoulder blades are on the sides of its chest, not on the back as ours are, and its collar bones are no more than ligaments connecting the breastbone to the shoulder blades. The chest itself is more rounded than that of a human.

Its flexible spine allows a cat to twist around obstructions – and to wash almost the whole of its body. The structure of its shoulders allows it greater freedom of movement in its front legs and a long stride. That same structure, combined with the rounded chest, also allows the cat to bring its front feet beneath its body and walk safely along a narrow line. That is how it walks along walls without falling.

Friend or foe?

Diary · Five months

It was the height of the kitten season and as usual in the summer a succession of orphans and unwanted kittens came to us for rehoming. Fortunately few stayed long; kind people wanting kittens knew that we have rescue kittens wanting homes.

Up to the age of about ten weeks most kittens seem to be naturally friendly towards others of their own kind. But by about twelve weeks they have become suspicious of strangers. The older a kitten becomes the longer it takes for it to make friends with another cat. Adult cats may take weeks to become friends with another cat; indeed some that have been used to living on their own never do. Snorkel's kittens are now five months old and reluctant to make friends straight off even with a small kitten.

Whisky, eight weeks old, comes boldly to meet Septimus, giving a friendly tail-up signal. He sniffs Sep's nose while looking into his face but Sep is suspicious; he draws back, tail down, eyes half-closed.

Fred is also being most unfriendly to Whisky, glaring at him and making him feel decidedly unwelcome. Whisky shows his discomfort by avoiding eye contact with Fred.

Lucifer reinforces his unfriendly glare by baring his teeth and hissing. No blow is struck. Whisky again avoids eye contact and prepares to creep away. In a few days he and Snorkel's kittens had got to know one another and played happily together.

The kittens at six months

Almost six months old and the kittens are young cats. They can take care of themselves now, provided they are not simply abandoned. They still depend on humans for shelter, medical care, and their basic food. Cats can and do survive in the wild without human help, but the life is harsh and usually short. Most house cats that leave home soon die from hunger, cold, disease, or some combination of them.

Jane was anxious that this fate should not befall the family. A home has been found for Ferdinand and he has gone to live there.

The pictures of Fred and Barley show that the sexual organs are well developed, but they are not yet fully mature. The average male is able to mate when he is about forty-seven weeks old. Septimus appears to be less well developed; he was a 'cryptorchid' with only one testicle descended. It is now time for all the male kittens to be neutered. For most of them the operation was simple but Septimus needs a slightly more complicated abdominal operation.

Septimus

Fred

Lucifer

Ginger

Tabitha

Barley

Staking a claim

The cats had special places in their home range where they often performed the same stereotyped actions each time they passed. A certain dead branch was ideal for claw-sharpening, for example. To photograph the cats performing their reactions at each spot it was only necessary to take them to it.

Opposite page Septimus sharpening *his claws. This removes the protective horny sheath that grows over the claws, so exposing the sharp tips. At the same time it deposits a scent mark on the wood. Claw-sharpening is also a comforting activity, which cats seem to enjoy for its own sake. The action is almost identical to that of milk-treading. Cats that walk the streets or on concrete surfaces wear down claws that grow too long, but a cat that spends most of its time indoors may need its claws clipped – using ordinary nail clippers, but being careful not to cut the quick, which is clearly visible. It is a good idea to provide a piece of wood, perhaps with a bit of old carpet nailed to it, for claw-sharpening if the cat lives mainly indoors.*

A territorial animal is one that occupies a particular area which it defends against all intruders of the same species. Many birds are territorial and constant vigilance is the price they must pay for exclusive access to food and shelter. They spend a great deal of their time patrolling their boundaries and singing to announce their presence. It is hard work. Cats are not hard workers, and in the strict sense they are not territorial.

A cat has a 'range'. This is the area within which it hunts for food and it may share its range with other cats. Females especially have ranges that overlap. Males have larger ranges than females and the boundaries of a male range ignore boundaries of female ranges. In any particular area, therefore, the mice may have to watch out for one or more hunting females and for at least one male.

Somewhere deep inside its range a cat will have a 'core area'. This is the nearest it gets to a true territory. It is the place where it knows it is safe. It can shelter there, sleep without fear of disturbance, take food to be eaten in peace, and for female cats it is in a nest located inside the core area that her kittens are born. A cat may defend its core area but not its range. This may not mean a fight. Unless the trespasser is younger, healthier, stronger, and determined to depose the occupant it will usually feel disadvantaged and depart.

House cats regard the house itself, and sometimes a small area around it, as the core area, but they wander much further in search of food. In rural England the average area for a cat's range is about 20-25 acres (8-10 hectares), but this is very much an average and in parts of North America ranges are up to ten times bigger than this. It has been estimated that if Britain were wholly rural it could support a population of around 2.25 million cats living entirely by hunting and there may be close to two million farm cats and cats living wild in Britain. If pet cats are included the population may be seven million.

Be that as it may, a male has a range roughly ten times larger than a female, and in towns the size of a range depends very much on the amount of food the cat receives from humans. The less it needs to hunt, the smaller its range will be.

Flags and messages

We think of a range as an area, but it would be more accurate to describe it as a network. Like most animals the cat goes about its business in a methodical way and once it has established a routine it does not waste time or look for trouble by venturing along unknown paths.

The cat does not patrol the boundaries of its range, does not defend its feeding ground, but it does announce its presence for the benefit of other cats. Along any path there are landmarks, points that you or I might overlook, but which seem prominent to a cat, or which have been used from time immemorial as

flagpoles or message boxes. Each time the cat passes a landmark it will inspect it to see whether other cats have passed that way recently, and then it will leave a message of its own.

The message is written in scent secreted by the cat's own glands and there are several ways to deposit it. Scratching, or claw-sharpening, leaves scent from the paws, and if the flagpole is suitable and the claws need to be sharpened, or the cat feels relaxed it may scratch. It may rub its face against a stone, tree trunk, or fence post (see page 107), leaving scent from its facial glands. Or it may spray.

Males spray more frequently than females, but females can and do spray from time to time. The scent is powerful and lasts a long time, so the message is enduring, more like painting on the wall than scribbling in chalk. It is urine that is sprayed, but the action is not that of urination, and most probably it is not ordinary urine.

The spraying cat stands upright and sprays horizontally. If it were urinating it would crouch and the liquid would fall downwards. Only a very small quantity of liquid is used – much less than when the cat urinates – and the urine itself is either

Septimus is biting a bramble stem and rubbing the side of his mouth and face on it. This leaves his scent on top of the probably fading scent left there by the last cat to pass.

highly concentrated or 'doped' with secretions from scent glands in the anal region.

When a cat finds a place that has been sprayed it may respond with what is called the 'flehmen' reaction. It touches the sprayed spot with its nose, lays back its ears, makes licking movements across the roof of its mouth, then raises its head with its mouth half open and its upper lip drawn back and stands absolutely still, staring to the front and breathing slowly. Inside the roof of its mouth a cat has a 'Jacobson's organ', very sensitive to chemical stimuli it receives from the tip of the tongue. The organ is most highly developed in reptiles – the snake flicks its tongue to capture molecules which it takes to its Jacobson's organ for examination – but many mammals, although not humans, also have it. The cat exhibiting the flehmen response is using its Jacobson's organ.

In some animals, spraying and the flehmen response to it are a form of sexual communication. This may also be true among cats, but no one is certain. It seems that spraying conveys the simple message that a cat has passed. If the scent is recent it is left, but if it is fading the cat that finds it must renew it.

Fred stands erect, tail a-quiver, as he goes through the actions of spraying. He has not yet mastered the technique of the full spray.

Barley and Ginger passing one another among the bracken, with a brief sniff and greeting, just to establish identities.

Diary · Eight months

The young cats prowled the garden and immediate woods and heathland, usually singly or in twos. The autumn woods were full of the summer crop of young mice and voles, and a happy hunting ground for cats, even for those that did not have to earn their whole living at hunting.

After a few minutes spent studying the scent left on some dead grasses by a passing stranger, Tabitha gives a flehmen response.

Sometimes cats will leave faeces unburied as scent markers. Many mammals do this, including tigers, but in cats it has a special significance. It is only dominant males that do it, and usually the droppings are left in a conspicuous place beside a path near the edge of a disputed boundary, for although ranges are not seriously defended, the unrestricted right of a male to 'his' females may be guarded and a rival may be unwelcome.

It is when you watch a cat moving about its range, stopping every few paces to sniff, that you appreciate the importance to it of scent, and messages conveyed by scent. Now watch the behaviour of a cat in the house, among humans, and you will see that scent is no less important indoors.

The cat that rubs its face against your legs is greeting you and marking you at the same time. You now carry its scent, which means you are part of its social group, and not a stranger. If you wash the cushion covers or leave towels or woollens where the cat can lie on them, before it curls up to sleep it is likely to tread them. This is comforting to the cat, but it also deposits its scent to mark the place as its own.

When you have stroked the cat, or played with it, the cat may wash itself thoroughly. This, too, is partly connected with scent communication. You, too, leave a scent on the cat when you handle it. By licking its fur the cat can check on your smell, reinforce its own smell, and groom itself all at the same time.

Living off the land

Cats are among the most specialized of carnivores. They live by hunting, and there is strength, grace and beauty in their movements, and careful premeditation in the slow approach that precedes the final, devastating attack. The obvious facts about the way they live have given rise to several myths about them.

Because nature has equipped them with all the weapons they need to pursue their life as hunters, many people suppose they are born with the skills they need for hunting and there is nothing for them to learn. The truth is that the young cat has much to learn.

The fact that domestic cats rarely share their food with one another has encouraged the view that they are solitary, anti-social animals. They are not.

Because they hunt efficiently some people believe they are cruel, ruthless killers. They are not.

Among cats that live wild, learning begins as soon as the kittens are weaned. Their first solid food will have been prey, caught by their mother and presented to them dead. Once they

A morning in early autumn and Barley, heading for home after a routine prowl, pauses and looks back, one paw still raised, as a small sound catches his attention.

have learned to recognize what is edible, the mother will start bringing home prey that is still alive and will kill it in front of them. Later she will take the kittens with her on hunting expeditions so they can watch. They will imitate her actions, clumsily at first, but with growing confidence and precision. When they start hunting by themselves they will use the 'orthodox' methods of attack, but then they may try a few experiments. Instead of grabbing a mouse by the back of the neck, for example, a young cat might grab at its flank – and be bitten on the nose as the mouse twists around. So, little by little, the cat learns there is a good reason for everything.

A cat that was not taught to hunt by an older cat may still turn into a hunter, but it will never hunt as efficiently, or as much, as a cat that has been properly 'educated'. The inclination to hunt is innate in the cat, and the sight of prey may stimulate it, but technique must be learned and improved.

The technique of the hunter

As the cat moves along the tracks that lace its range it is constantly alert. The slightest rustle of a leaf will make it pause, the lightest breeze from an unexpected quarter will be investigated. If the cat hears or sees a suitable small animal it may well begin to stalk it. Sometimes part of the range is open, for example in a field from which a crop has just been harvested, and you can watch the cat patrolling slowly, deliberately, back and forth by invisible paths.

Domestic cats usually hunt alone and eat what they kill themselves rather than sharing food as do some of the larger predators, but there are exceptions. The sharing of food is not unknown, and sometimes two or three cats will hunt in the same area together. It is not certain that they are collaborating intentionally, but cats have been observed working their way across a field in an open kind of formation so that by running away from one cat prey would be likely to be caught by another. More generally, though, cats hunt alone because it is more efficient when the prey is very small and very nervous. A cat needs no help to kill a mouse and there is not much of it to share. The hunting technique is related to the kind of prey and not to the social organization of the cats. Because they usually hunt alone it does not follow that they live alone.

The hunter will take advantage of any opportunity, but most prey is not caught in the open. The prowling cat is not even looking for prey so much as for places frequented by prey. It will notice the entrances to mouse or vole burrows, the tracks made by the residents of those burrows, droppings, the small platforms where seeds and berries are eaten, and when it finds such a place the cat will sit awhile and wait. As it waits by its mousehole the cat looks half asleep. It looks to its front – not directly at the hole – and remains motionless.

Diary · Eight months

The six young cats varied a great deal in the amount of hunting they did. Perhaps some were more clever at it, while others were just lazy and too well fed to bother. Septimus was the greatest hunter. He would pounce on anything that moved: flies sunning themselves, butterflies feeding at flowers, grasshoppers, beetles, mice. Tabitha was also very alert and active. Ginger preferred the quiet life, staying at home, curled up or stretched out in the sun.

Right top *Prowling cats cannot see far because most plants are taller than they are. Septimus stands up to see over the top of the long grass.* Right bottom *While stalking, the cat moves forward slowly, cautiously, and as inconspicuously as it can. If there were cover, Septimus might use it.* Opposite page *The mouse is dead but Septimus plays with it as he used to play with prey brought in by Snorkel when he was a kitten.*

Cats hate to be cold or wet, and snow brings misery. If his fur gets wet Septimus will lose body heat and chill rapidly – and he knows it. His fur is the only protection he has against cold and he must keep it dry at all costs. He takes refuge in the same woodpile he used to keep cool in the summer (see page 61).

The peripheral vision of a cat – the awareness of the smallest movement at the very edge of its visual field – is acute, and it is for that small movement that it waits. When it happens, and a mouse appears, the cat strikes.

Unless something goes wrong, the kill is instantaneous, administered by a bite that severs the spinal column between two neck vertebrae. It is not cruel. The mouse suffers no pain and may not have time even to suffer fear, and its death is necessary, for the cat kills that it may eat.

Cats rely mainly on mice and voles, but they are more versatile than this makes them seem. They kill shrews, possibly mistaking them for mice, but rarely eat them and often vomit when they do, probably because shrews secrete a poison that makes them distasteful. They will kill young or sick rabbits, but a healthy adult rabbit is too large and strong for most cats to overpower. Some brave cats will kill rats, and squirrels and even weasels and stoats. Many cats eat insects, some eat worms, which can be large and highly nutritious but which anchor themselves firmly to the mouth of their burrows while they feed on the surface and are ready to retreat very fast. Some cats eat reptiles, some fish, and if they find a bat roost they will even take sleeping bats.

So far as cats are concerned, birds cheat. On the ground a small bird looks and moves much like a mouse. The cat will stalk it, but as it pounces, expecting the prey either to remain where it is or to run along the ground, the bird jumps into the air and stays there. It can see behind itself, which complicates matters further. Some cats learn to pounce upward, to land on top of the bird just as it takes off, but many cats have to make do with fledglings that have fluttered from their nests and have not yet learned to fly from danger. Catching birds is difficult enough, but to make things more difficult, some birds are superb ventriloquists. They utter warning cries, but it is impossible even for a cat to locate the source of those cries. Sometimes you will hear a cat chattering in frustration, or just in wishful thinking, at birds it can see but not eat.

The weapons

When stalking, the cat moves very slowly, uses any available cover. When it is within range it draws its hind legs forward beneath its body, their claws giving purchase like the spiked shoes of a sprinter, and as the hunter leaps it stretches its body and brings the front legs forward in a single, huge stride.

The running cat extends its front and hind legs fully so that the whole cat is at full stretch during each bound (see pages 98-99). It is 'digitigrade' – it walks on its toes rather than on the soles of its feet as we do. This increases its speed. By flexing its supple spine to bring its back legs forward for the next leap and propelling itself with its front legs as well, the cat can outrun most of the animals it meets in its daily life. It cannot keep up this pace for long, however, because its chest cavity is small and its heart and lungs cannot supply its muscles with enough oxygen for prolonged effort. It tires quickly.

Once the prey is caught the front claws are used to hold it while the canine teeth feel for two neck vertebrae and prise them apart. The canines are very sensitive to touch and though large they are not strong, and break easily. They are precision tools, as are the tiny incisors, used mainly for grooming. Cats cannot chew. Their molars are long, sharp ridges called 'carnassials', teeth with cutting edges that work like scissors to cut food into pieces small enough to be swallowed.

Eyes in the dark

Once the kittens' eyes attained their adult colouring I often noticed that in certain lights Ginger's eyes glowed a copper colour, whereas the other's eyes, especially those of the tabbies, reflected green or white, depending on the brightness of the reflection. We photographed Ginger's eyes at night to try and record this ginger glow, using a technique combining ring-flash with back-lighting.

Above *and* opposite page *In some photographs Ginger appeared to have eyes like opals, glowing with flecks of different colours within them.*

A cat's eyes are very large in relation to the rest of its face. It is one of the features we find attractive, but large eyes are an adaptation to hunting by night and so most nocturnal animals have them. The reason is simple: the larger the eye the more light it will admit, but there are more differences than that.

Diurnal animals, including humans, have an indentation, the 'fovea', in the centre of the retina which helps focus the image the eye receives. Along with other nocturnal species, cats have no fovea. Nor is the lens of the cat's eye able to change its shape and so focus light as well as that of the human eye. This means that what a cat sees is less sharp than what we see, and they see a sharper image at the edge of their field of vision than they do at the centre. This is why they spend so much of their time apparently staring into space with their eyes unfocused, but it is not so bad as it seems. Our brains, and presumably those of cats, too, interpret what we see so we have the impression of images which are considerably sharper than those that our eyes actually transmit.

The eye of a cat admits about half as much light again as does the human eye, and from a larger area, so the cat has a wider field of vision than a human. The retina, at the back of the eye, is where the light is received by two kinds of cell, rods and cones. Rods are sensitive to any light, and cones discriminate colour. The human retina has about four times as many rods as cones. The cat's retina has about twenty rods to each cone, so its colour vision is inferior to ours, but what it sees is about five times brighter than the image a human would see under similar conditions, so it sees much better in dim light. The cat cannot literally see in the dark, but it can see in what we would think of as total darkness.

Night shine

With eyes as sensitive as this, the cat must limit the amount of light that enters and in bright light its iris closes to a narrow slit, rather than the pinprick of the human eye. This vertical pupil, which is found in many nocturnal animals but not in diurnal ones – not even diurnal cats, such as the lion – seems to give finer tuning.

The retina of the cat's eye also has a small triangular patch of cells containing up to ten layers of crystals made from zinc and protein, called the 'tapetum lucidum'. The tapetum is like a mirror, reflecting on to the retina incoming light that has not been absorbed. The protein may be, or may contain, riboflavin, which absorbs light and emits it again at a longer wavelength – it fluoresces. At night the pupil is opened to its fullest extent. If the cat turns at a sudden noise, and if the noise is accompanied by a bright light such as a torch or the headlights of a car, there will be a pause before the pupils contract during which light is reflected by the tapetum. That is 'night shine'.

Mealtimes

Our cats still had quite nice manners – they did not growl or lash out at each other if eating from one dish. But I had to feed them separately because there were one or two greedy guzzlers among them that simply bolted the food down faster than anyone else. Ginger got quite stout gobbling more than his fair share, while Septimus became almost skinny, because he was such a slow feeder. By now Lucifer and Barley had gone to new homes and so we had only four of Snorkel's kittens left.

Cats are carnivores. They eat meat, and if they are to remain healthy, meat is what they must be given and they need a fairly large amount of it in relation to their body size. Their carnassial teeth are designed for cutting, not grinding, and their guts are short and supplied with strong acids to digest meat but not raw vegetable foods. However, cats do chew grass which supplies them with the folic acid they cannot obtain from meat. Cats can digest cooked root vegetables and some other plant foods such as cooked rice, bread, cake and biscuits.

If it lives wild a cat will feed mainly on small rodents, and provided there are enough of them and it is good enough at catching them, the cat that eats nothing but mice will be well nourished and healthy. In preparing cat food it makes sense to find out what mice are made of and try to match that composition. Mouse is about 70 per cent water, 15 per cent protein, 10 per cent fat, and one per cent carbohydrate. The moisture content is important, as cats obtain much of the liquid they need from their food. Fresh meat contains about 70 per cent water, fish 80 per cent, and canned cat food 75 per cent.

In addition to the major nutrients, mice provide the cat with vitamins A and those of the B group, and with calcium and phosphororus, for the cat eats the whole mouse, bones and all.

1 *Ginger approaches his bowl with his tail up. As he starts feeding he closes his eyes, then lowers his tail, and sinks into a crouch.*

2 *As Fred feeds he opens his eyes occasionally and looks around, to ensure the others are getting on with their own dinners and are not after his.*

An adult female cat needs 200-250 Calories a day, an adult male 250-300, and a growing kitten more in relation to its body weight than an adult. One pint of milk provides 400 Calories (700 Calories per litre) and meat or fish provide about 500-600 Calories per pound (1100-1300 per kilogram), the less fat in the meat the fewer the Calories. Meat can be fed raw or cooked, but if it is cooked some vitamins are lost, so it should be cooked only lightly. Too much liver or cod liver oil will supply an excess of vitamin A which causes bony growths around joints. Most cats will drink milk, but some are allergic to cow's milk and others cannot digest lactose so milk upsets their digestion.

Proprietary canned foods are convenient and guaranteed to provide a complete and wholesome diet. Dried foods also provide a complete diet, but with a very low moisture content. A kitten weaned on to dried food will usually drink enough water to keep it healthy but an older cat unfamiliar with dried food may drink too little and suffer dehydration. Adding a little salt to food will encourage a cat to drink more, and clean, fresh water must be available at all times.

In the wild, a cat will gorge itself occasionally and then rest. Domestic cats are used to a more regular regime of one, or more commonly two, meals a day. Food should be room temperature.

3 *His bowl licked clean, and working by scent rather than sight, Septimus sniffs carefully all around it to check that no tiny crumbs have been overlooked. Most cats are neat feeders. Food that needs cutting into smaller pieces will be lifted from bowl to floor, then eaten before the bowl is examined again.*

Grooming

When all the bowls had been licked clean each cat moved away from its own bowl, and perhaps stretched as it went. A quick inspection of everyone else's bowl reavealed little in the way of leftovers, but they were all sniffed over and licked again. Each of the cats then moved off a little way, sat down, and groomed itself.

The cat spends a great deal of time caring for its fur and although 'allogrooming' – the grooming of one cat by another – is a mark of affection, the serious business of fur maintenance must be attended to by each cat for itself – as 'autogrooming'. The cat grooms itself when it feels the need, but also as a displacement activity when it is confused, or needs a moment to decide what to do next. Cats will groom themselves while they work out solutions to such important problems as how to get at the milk without being caught.

Three sets of tools are used for grooming – the tongue, the incisor teeth and the front paws, and the fur itself helps. Each hair is covered in minute scales that catch on the scales of adjacent hairs, but the hairs move against one another in only one direction, like a ratchet. If a hair is out of place anything that moves it will tend to restore it to its correct position.

The rough tongue is covered in hard, flexible spines pointing backwards, towards the throat. They help rasp fragments of food, but they also make the tongue into a useful brush. The incisor teeth are tiny, but excellent for extracting flakes of dead skin, bits of matted fur and bits of grit. The paws are used to carry saliva to those parts of the body – such as the top and back of the head – where the mouth cannot reach. There are not many such

1 *Following a feed, Ginger is polishing his muzzle, licking all around it as far as his tongue will reach. This lip-licking follows all meals.*

2 *Fred licks his left front paw thoroughly and when he is satisfied with it he rubs it repeatedly over his face, licking and rubbing alternately.*

places, because the cat has such a flexible spine.

A cat that spends much of its time indoors benefits from regular brushing, especially if it is long-haired. Outside many of its loose hairs would be caught on vegetation or just blown away, but indoors they have to be removed by grooming, and apart from those littering the furniture they are swallowed. Fur balls lodged in the gut are often removed by vomiting, but they may also cause loss of appetite leading to weight loss, other digestive complaints, and even respiratory complaints.

If the cat spends much time outdoors it is almost certain to collect fleas and bring them into the house. Cat fleas are a different species from human fleas and transmit no diseases, but if there is no other host within reach they will bite a human. They hop on to a passing host, feed on the blood, then fall to the ground. This means there are ten or more times the number of fleas in places the infested cat frequents than on the cat itself. Applying an insecticide to the cat has little effect. The fleas should be attacked by vacuuming indoors and spraying places the cat visits outdoors. Fortunately the problem is self-limiting because the fleas are subtropical animals and do not survive for long in cool, damp conditions. A spell of warm dry weather will allow them to breed, but in cold weather most of them will die.

3 *Ginger is washing behind his ears. He rubs his licked paw from his ear over his eye and down past his whiskers many times, licking the paw repeatedly.*

4 *Tabitha has turned her head through nearly 180 degrees and the spines on her tongue are smoothing the live hairs and removing dead ones.*

The cats in action

Cats can leap several times their own body length either horizontally or vertically, and can put on a fair turn of speed on the ground. Multiple-image photography has been used to show how they accomplish some of these feats. This requires complicated and not very portable equipment, so the cats had to be trained to perform in the right places. The training was done when, apart from young Ferdinand opposite, the cats were nearly mature and their strength and agility fully developed. They were presented with obstacles and situations they might meet in everyday life and encouraged, sometimes with a crumb of food, to overcome them. They seemed to find the experience pleasurable, and Fred soon emerged as a fast learner.

Falling

Cats are nimble climbers and rarely fall, but if they do they always land on their feet. This remarkable ability has long been a source of puzzlement because it apparently defies one of the fundamental laws of motion, which states that action and reaction are equal: the force required to turn an object around must have something to push against. A free-falling cat clearly has nothing to push against, so how does it turn? At first people thought the cat's tail acted as a sort of propeller using air resistance to turn the body, but these pictures show that the tail does not take an active part. In fact, cats use another law of mechanics – the one skaters use when they increase the speed of their turn by withdrawing their outstretched arms. Note how the young Ferdinand twists the front half of his body, paws tucked in, against the rear half, with paws outstretched (image 4, side view). By image 6 he is doing the reverse. His hind legs are tucked in and his forelegs are outstretched. In this way movement of the forequarters is transferred to the hind quarters, and the whole body is turned. Young cats acquire this ability at a very early age and can turn in a remarkably short distance. In these pictures Ferdinand has been trained to fall on to a soft cushion and was taking his time in turning.

Taking off

The first image shows Fred crouching
ready to spring with his eyes fixed on
the shelf where he will alight. In image
2 his body is starting to extend but his
hind feet have not moved and are still
flat on the floor. The hind legs then
straighten with the body fully extended
and the tail brought over the back. In
this position the tail is a sort of energy
reserve, carried over the back until his
front paws touch the shelf, when it is
whisked down, giving his hindquarters
a small boost.

Landing

A cat descends headfirst with its body fully extended. On impact, the body is doubled up, with the hind feet brought close to or even in front of the forefeet, as shown here. In this way the shock of landing is spread over all four feet, and over as long a period as possible. If all four feet touched down simultaneously the shock would be considerably greater.

Jumping off a shelf

Jumping off a shelf requires a cat to launch itself directly into the air. Once its hind feet break contact, it has only its tail and sense of balance to guide it. Forepaws are initially tucked softly beneath the chin, but as the ground rapidly approaches, the forelimbs are fully stretched and the toes spread. The wrists flex considerably as the fore limbs take the initial shock of landing, but the hind limbs are quickly brought down to take their share of the impact.

Jumping down a wall

A cat can make use of the vertical surface when it jumps down from a wall. By extending its body and forelimbs as far as possible down the wall before letting go it can effectively reduce the height from which it jumps. Here Ginger seems to run down the wall for more than a body-length before pushing off to land neatly a short way from its base. The push-off is necessary to provide room for him to bring down his hindquarters. Compare this with the way a cat climbs down a ladder, on page 66.

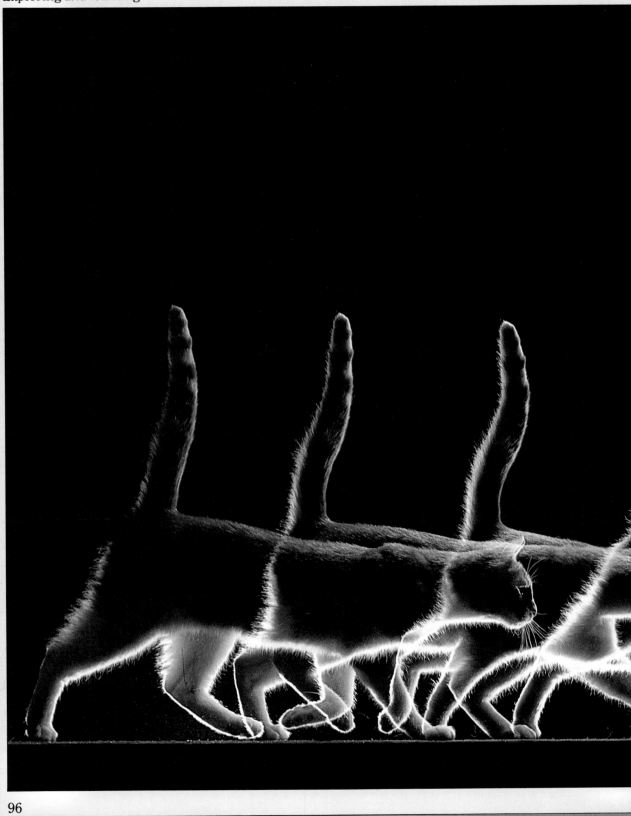

Trotting

When an animal trots, two feet are always on the ground. While the fore-right and hind-left feet are moved, the other two are still in contact with the ground, and vice versa. In the first image Fred is seen with both his left feet off the ground. In fact he is about to put down his left hind foot and raise his right hind foot to move forward together with his left forefoot. Cats trot for short periods only, unlike dogs which can trot for miles. Here Fred is trotting to receive a titbit, and shows his pleasure by holding his tail aloft.

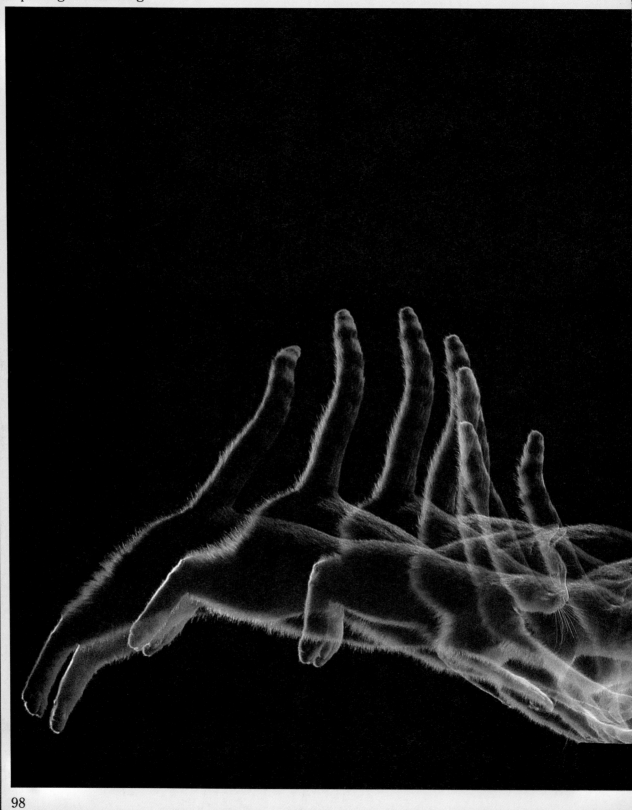

Bounding

A bounding or galloping animal progresses in a series of leaps, losing all contact with the ground during each stride. This is the fastest means of locomotion and cats generally do it only in play or to escape from extreme danger, or when capturing prey. Bounding involves the use of the whole body, not just the legs. Much of the energy needed comes from the back muscles which flex and extend the spine during each leap. Here Fred is bounding with his tail up. This is a play situation; if he were escaping his tail would be hooked.

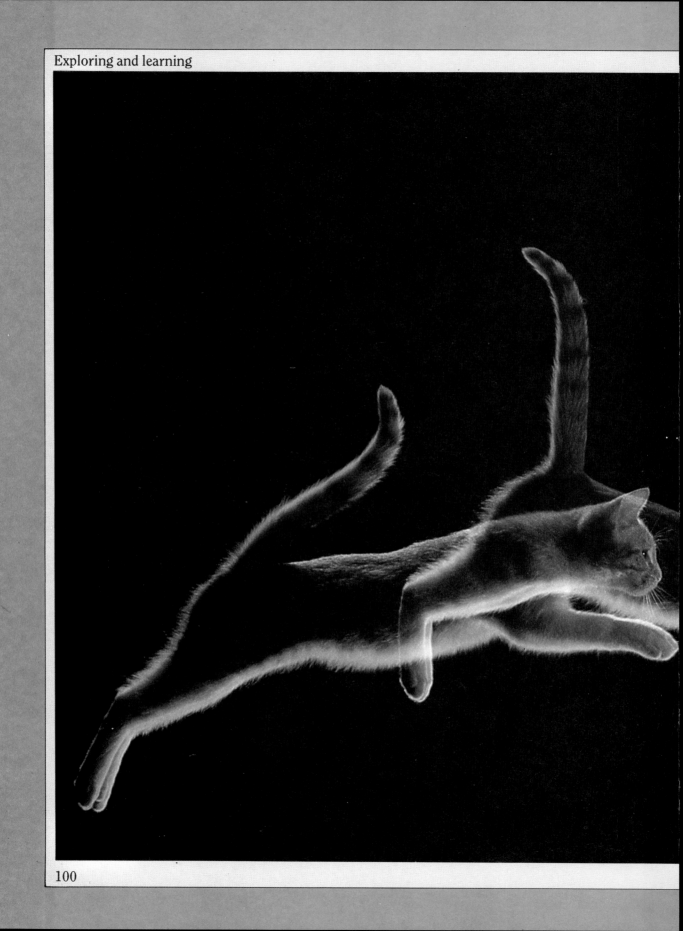

Leaping a gap

All the movements of cats appear to be carefully considered. Cats hardly ever take a running jump the way dogs do. A gap to be jumped is sized carefully for some time and the firmness of the substrate tested with the hind feet before the spring is made. Here Fred, a nearly mature cat, has judged to a fine degree his own ability to clear a gap. Note how the tail is used to give a boost at the end of the jump. But judging a jump has to be learned – as Septimus discovered to his cost when he tried to leap over water (see page 58).

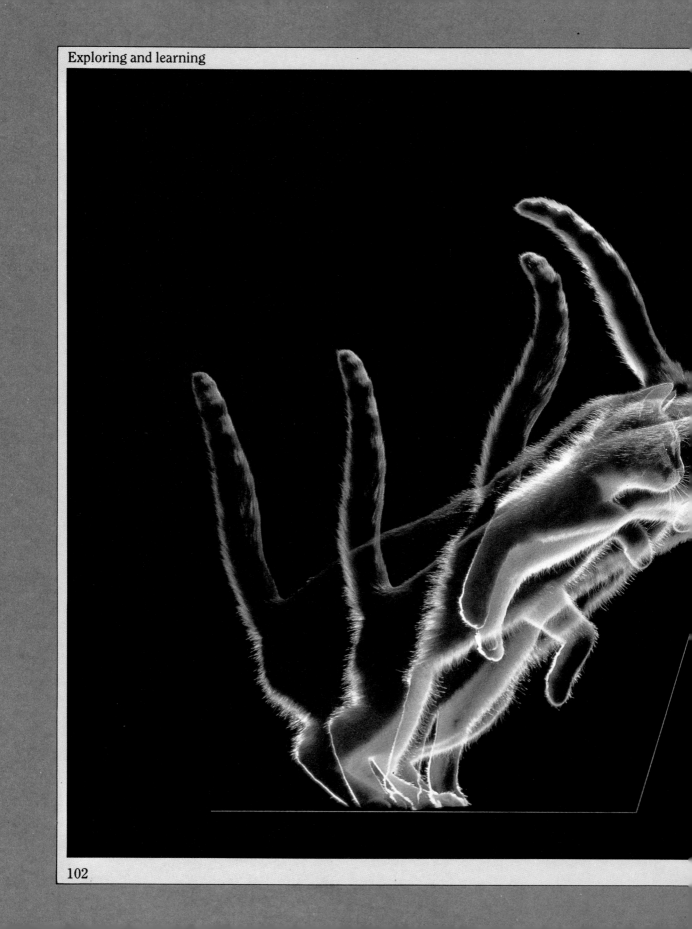

Jumping over

Cats are careful and precise in their movements. A minimum of energy is expended to accomplish a desired end. Here Fred discovers a low obstacle in his path. Instead of scrambling over it, he performs a beautifully controlled action which carries him to the other side without touching the obstacle. His judgement is practically to within a hair's breadth.

Coming of age

Ten months had passed since the kittens were born. All except for Ginger and Tabitha had left, and they were kittens no longer but cats capable of finding their own food, of patrolling the range each of them had established, of taking care of themselves. Very soon Tabitha would be old enough to prepare a nest and care for kittens of her own.

She was entering puberty, and becoming aware of her own sexuality, driven by a hormonal system that was being 'switched on' a little at a time, to behave in ways calculated to attract a mate before her body started to secrete the chemical attractants whose scent would capture and bring him to her. This was the first stage in her regular sexual cycle and might occur more than once before the second stage followed it and she mated and became pregnant.

Tabitha rolls playfully as she luxuriates in the warm spring sunshine. At ten months old she is just starting to call, and to emit the scents that will attract all the neighbouring unneutered toms.

Grown up and dating

Just as he did when Snorkel came on heat, Fergus materialized out of the woods as soon as Tabitha started calling. Other tom cats were around: Mr Tibbs from next door no doubt, and a big black male called Bertie.

All female mammals go through a regular reproductive cycle called the 'oestrus' cycle. There are several versions and cats, like humans, experience 'polyoestrus'. This means they are sexually receptive and able to become pregnant more than once during a year. Female cats are receptive to males only when they are ready to be fertilized, and instead of being monthly like that of humans, their cycle repeats itself once a fortnight. The more important difference between the reproductive method of cats and humans concerns the ovum – the egg. In humans an ovum is released during each cycle. It enters the uterus, waits there to be fertilized, and if it is not fertilized it is shed. The cat does not release ova until after mating. It is the act of mating that stimulates ovulation.

This is how cats have adapted evolutionarily to a life in which the two sexes live separately. The female cannot be sure that a male will be in the neighbourhood when she is ready to receive him, so she must give each mating the greatest possible chance of leading to pregnancy. The method has the advantage of conserving ova, since if for any reason the cat remains unmated she will not ovulate.

Rolling and calling

The method also affects the way the female cat behaves. She must attract a male when she is ready to release ova, and she does so very insistently. While she is unreceptive she will engage in no sexual activity whatever. This phase is called 'anoestrus'. Her period of receptiveness, oestrus itself, is preceded by 'pro-oestrus,' lasting for about two days. During pro-oestrus she will become increasingly affectionate. She will rub herself against objects – or people. She will mark a particular spot on the ground and rub her chin on it, eventually writhing ecstatically, rolling, and crouching close to the ground with her hindquarters raised and her tail held to one side. She will call, with an unmistakable loud – and in Siamese cats a deafening – persistent raucous cry, and as she passes from the pro-oestrus stage in to the oestrus phase, she will start to accept the advances of the tom cat. Initially she flirts outrageously with him, luring him, then rebuffing him; but finally she allows him close enough first to sniff her, then to grab her by the scruff and finally to mount her.

If the female mates, oestrus usually continues for a further two or four days after the mating. If she does not mate it will last for nine or ten days. If she becomes pregnant the hormonal system adjusts to the new state and she will not call again until a few weeks after the kittens have been weaned, the first full oestrus sometimes being preceded a week or two earlier by a 'false oestrus' – a pro-oestrus with no oestrus to follow it – as it was with Snorkel. An unmated female will start an oestrus cycle about two weeks after the last one ended.

Should she be neutered?

If you own a female cat you have three alternatives. You can allow the cat to mate, provided there are intact males around – and these days intact males are not so common as they were – and provided you can find homes for the kittens or are prepared to keep them yourself. Or you can prevent the cat from mating, by keeping her indoors perhaps, if you are prepared to endure her calling, and her air of desperation as she begins to tire and still no mate appears, twice a month. The third alternative is to have her neutered.

It is not quite so simple as it seems because neutered females may still go through the first part of pro-oestrus, although it is unusual for them to reach the stage of calling.

Neutering, or 'spaying' involves surgery under a general anaesthetic and conception can be prevented without removing the entire reproductive tract. The ovaries only may be removed, or one can remove the ovaries together with the fallopian tubes and the uterus. If less than the full tract is removed it is quite possible for some residual hormonal activity to continue after the operation.

For some reason Fergus went into ecstasies over this fence post, chinning, rolling and repeatedly rubbing his old bullet head hard against it. Perhaps Tabitha had been this way.

In Europe, Australia and New Zealand it is usual to make the incision through the flank, after shaving a patch of skin which remains naked for a few weeks after the operation. In America the incision is usually made in the middle of the underside of the abdomen, where it is less obvious. The operation does not take long, is very routine and as safe as any surgery can be, and the cat taken to the veterinary surgery in the morning can generally be collected later the same day. She will have two or three stitches that may have to be removed about a week later.

Spaying is permanent and irreversible, so it is no good changing your mind later and deciding you would like your cat to have a litter. There are oral contraceptives for cats. They are used in some places to reduce the population of homeless town cats where it does not matter which cats have kittens so long as most of them do not. Contraception is reversible, of course, but drugs are unreliable unless administered regularly.

Does spaying make a cat fat and lazy?

If you prevent the secretion of the hormones that trigger sexual activity, obviously sexual activity will cease. To that extent a neutered cat behaves differently from one that is intact. Apart from that, the neutered cat is the same cat she was before the operation, with the same personality and the same likes and dislikes. Spaying does not make a cat fat and sluggish. It is over-feeding that makes it so.

Opposite page *and* left *Fergus, chinning delightedly at a fence post, hugging and climbing up it, then rolling on his back at its base. Perhaps he has smelled something similar to a sexual attractant. This is how many cats react to certain smells – such as catnip. It is a 'superstimulus' causing an exaggerated version of the way a female behaves in oestrus, but affects males and females, neutered and intact alike.*

Battling toms

When Fergus came in from the woods his son was taken aback. He had been gone so long Ginger had forgotten he was father and friend, and Fergus had forgotten that Ginger was his son and playmate. Ginger had been castrated at sexual maturity, so was not especially interested in Tabitha, nor did he pose any threat to Fergus.

Fergus asserts his dominance as he and Ginger meet. Ginger crouches submissively and nervously – his ears are flattened to the sides. Both are snarling.

Cats are not aggressive animals and rarely fight, but in towns, where the cats are more crowded than they might be in a rural area, the crowding can produce noisy conflict among males competing for the favours of a female. It is this conflict, and not some kind of amorous serenade, that we hear as caterwauling.

A male hunts within a range that includes the ranges of several females. When a female is in oestrus she leaves characteristic scent marks to attract the male, and calls to direct him to her. He arrives, mating follows, and the encounter is peaceful. In a town, however, cats cannot always choose where they live because their human owners provide them with the core area in which they rest and where they receive most of their food and the core area has to lie inside the hunting range. Male ranges overlap, and the ranges of cats living wild in the same area are superimposed on the ranges of the house cats. When a female calls, therefore, more than one male may arrive.

The presence of two males on the same quest requires them to decide between them which is entitled to mate. They do not think to consult the female to find out whether she has an opinion in the matter – although she does – but proceed at once to declare their ranks. This involves threatening one another and the threats begin vocally. The two combatants square up to one another, ears back, and scream abuse. This may be enough, because in most cases one cat will be near its own core area, close to its home, and it will feel more confident than the other, which is further from home and may feel it is trespassing. So it will crouch in a submissive posture and retire, snarling. If it

does not the dispute may have to be resolved by blows. These may draw blood and leave ears tattered, but in most cases the fights are less dangerous than they look and sound. The purpose is to establish dominance, not to cause injury.

Unfortunately, the noise may attract other males, so the affair becomes a general jockeying for social position and the volume of the feline chorus increases accordingly.

If a tom is neutered before he reaches full sexual maturity he will never be concerned about rivals. He will have no social status, but usually neutered toms lead happy lives and remain friendly and playful. If he has had sexual experience a tom may continue to pursue females, and even to mate with them, for some time after the operation, or even for the rest of his life.

Above *Ginger has a threatening expression and keeps up a constant 'yowling' song.* Left *If Fergus comes too close Ginger draws back and explodes with a kind of spitting scream of defiance, rage and fear, but, having asserted himself, Fergus left, lashing his tail. Quarrels are usually resolved without coming to blows.*

Courtship and consummation

When I knew Tabitha was coming on heat I kept her shut indoors, as I had planned an arranged marriage for her to Mr Tibbs from next door. (Mr Tibbs had an exceptionally good temperament, and was quite handsome besides). For the first two days Tabitha and Mr Tibbs got to know each other through the bars of the 'play-pen' in which Tabitha was confined.

If you have a female cat and you wish her to produce kittens you can simply leave her to her own devices and let nature take its course, or, like Jane, you can arrange a marriage. These days, when so many domestic cats are neutered, arranged marriages are much the more reliable method.

The cats need time to get to know one another. Mr Tibbs came to stay with Tabitha, but it is more usual to take a female cat to the male. Male cats are extremely selfconscious and it is very difficult to persuade them to mate in strange surroundings. Even with a female enticing him with every wile in her repertoire, a nervous male may lose interest.

No matter what the males on top of the wall may think, the female chooses her mate. If there are several contenders she may accept the advances of one of the 'losers', and if you introduce her to a male she may not agree with your choice.

The first step, therefore, is to allow the female to come into oestrus, and time her cycle carefully. Then you will be able to predict more or less when she will be receptive again. Take her to the male when she starts to call. Keep her confined near him, so that both cats can get to know one another. Then, when the female is fully receptive the two cats will know each other and like one another well enough for the encounter to be productive.

The courtship begins with the female calling, rolling and rubbing, and the approach of the male. He will advance cautiously and get as close to the female as she will allow, but she may run a little distance away, then start calling again. As he advances he will rub his head against places where she has paused, and he will make soft, gently chirping sounds. This courtship may go on for hours. He may try a different technique, advancing a few steps when she looks away, then stopping at once and pretending complete disinterest if she turns in his direction. He will not be allowed too near. All cats have a 'personal space' surrounding them which must not be invaded without invitation. When he is close enough she may slap him with a paw and then he will stop and sit quite still. There is nothing else he can do, for if he comes closer she will either attack him for invasion or she will flee and he will then lose her altogether.

Eventually he will be allowed to touch her, sniff her and, at last, to mount her as she adopts the 'lordosis' position, crouching with her tail held to one side and her rump raised. Getting into the right position is a lengthy business but copulation is quick. It ends with a cry from the female, who turns and snarls at the male. It is not entirely certain whether the cry is of pleasure or pain, although pain is more likely because it is caused by the scratching of large spines on the penis of the male. These spines, very like those on a cat's tongue, increase in size as the male matures and secretes more testosterone and diminish as he grows older and his testosterone levels fall. Ovulation follows about one day after copulation, and probably it is the penile barbs which stimulate hormonal responses leading to ovulation.

Courtship and copulation will be repeated many times, often over several days. If the male tires of her, she may mate with another male.

Cats have preferences and in order to mate each of them must allow the other into his or her 'personal space', but the courtship ritual and mating technique is also related directly to the way cats live in the wild. By calling loudly, but refusing to mate for some time after a male has appeared, the female may attract more than one male which widens her choice.

If it is true that the penile barbs stimulate ovulation, matings with males in the prime of life are more likely to be productive than matings with old males. Encounters between cats may be rare in some environments, and by mating repeatedly once they have met, the probability of their producing a litter is much increased.

Homosexuality is fairly common among young toms, but is replaced by heterosexual behaviour when females in oestrus are available. Female homosexuality also occurs among females in oestrus with no access to males.

Mr Tibbs has come too close and has invaded Tabitha's 'personal space' so she boxes his ear, but gently. He stays where he is.

Opposite page *Tabitha is in the 'lordosis' position. Mr Tibbs sniffs at her, before grabbing her by the scruff.* Left *Mr Tibbs seizes her gently, his mouth remaining partly open, by the loose skin at the back of her neck. This grip restrains her.* Below *Mr Tibbs treads with his hind feet which stimulates her to raise her rump further.*

Left *As he penetrates, still grabbing her neck, the scratching of his penile barbs makes Tabitha cry out. The stimulus will lead to ovulation and probably to fertilization.* Top *Immediately, she twists and pulls away from Mr. Tibbs, snarling.* Above *Mr Tibbs backs away out of paw's reach.*

Above *The sequence ended, Tabitha rolls voluptuously and stretches towards Mr Tibbs, who watches her.* Below left *The two cats then wash themselves thoroughly in the genital region. From time to time Tabitha* rolls again in ecstatic languor. Below centre *Now both cats sit a little way apart. There is a pause for about half an hour during which the two of them just sit, quite relaxed, sometimes glancing at one another*

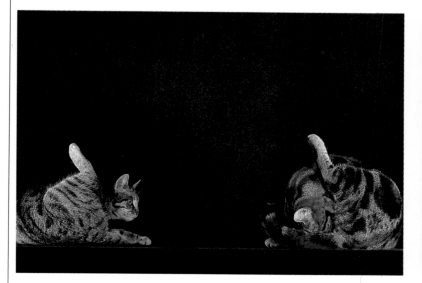

Diary · Mating

For two days Mr and Mrs Tibbs mated repeatedly, each time the pattern of behaviour being the same. Tabitha went off her food and hardly ate at all during this time; but Mr Tibbs, who had always had a hearty appetite, ate her dinner as well as his own. After three days, when mating had ceased, Mr Tibbs went back home. Tabitha rejoined her brothers and was exceptionally playful with them. They seemed rather staid old cats by comparison.

but mostly looking into the middle distance, ignoring one another.
Below right Then the courtship sequence begins again, led by Tabitha. To arouse his interest she edges closer to Mr Tibbs and taps

him playfully. As he approaches she will start rolling and moving away from him. The sequence is repeated many times. If Mr Tibbs was to tire first, then Tabitha would accept another male.

Full circle

Diary · Twelve months

Tabitha's kittens were due ten days after her first birthday. On the evening of the eighth day she ate her supper with no loss of appetite; there was no sign of restlessness or extra grooming or vaginal discharge which might indicate her kittens' early arrival, and she retired to her heated bed as usual.

But on the morning of the ninth day, I heard the mew of a newborn kitten even before I'd opened her bedroom door. I looked in her bed – nobody there. I checked all the alternative beds I had prepared – all empty! Then another mew, and I found the family in a bare corner behind a cupboard. One kitten was lying stretched out on the floorboards apparently dead; Tabitha was in a pool of wet, curled round a soggy, dark mass, but this mass was heaving, so some kittens were alive. There were four of them, all tabbies like the 'dead' one.

I put them in the heated bed; Tabitha quickly got in with them and curled herself around them. They would soon dry off in the warmth.

The fifth kitten, a female, was almost dry, but cold and stiff and not breathing. However, it is always worth trying to revive an apparently dead kitten, so I immersed it up to its ears in cat-temperature water for instant all-round warmth. Moments later it gave a gulp, then it curled into the kitten-being-carried posture. Soon it was breathing again. Its limbs began twitching, then paddling quite strongly.

I rubbed it vigorously and put it on a hot water bottle, wrapped in a dry towel. After about ten minutes, this kitten was clawing itself about, making hopeful sucking noises. I took it back to Tabitha and put it among the other kittens. An hour later I couldn't tell which had been the 'dead' kitten – all were dry, all tabbies and all females!

Five days later, all five kittens are doing well, their eyes just opening. When they have been weaned and found homes, Tabitha will be spayed. The kittens too, will be spayed when old enough. It would not be desirable to allow another thirty or more kittens to appear next year!

Tabitha on her first birthday. Five kitten embryos are shown in x-ray.

Tabitha lies happily suckling her five one-day-old kittens.

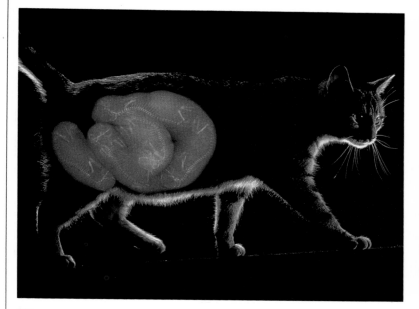

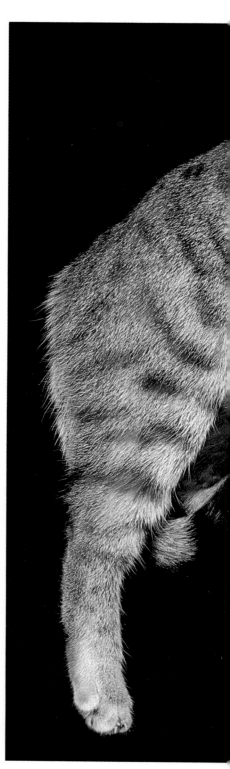

From birth to maturity-a summary

We end with a brief and very approximate guide to the rate at which our kittens matured, and invite you to record the development of your own kittens – but with a qualification. No two cats are alike. If a kitten reaches a particular stage earlier or later than another this does not mean it is better, or more or less healthy.

The beginning

Day 0
The kittens were born. The birth lasted for about twelve hours and produced seven kittens. Each kitten started breathing before the umbilicus was severed, and was immediately active. Newborn kittens are blind and deaf but can find a teat by touch and smell and start feeding almost at once. They can distinguish their own mother from other cats.

Day 4
The kittens are sleeping much of the time, and probably dreaming, but feeding strongly. They are starting 'milk treading.' They hiss at strange smells within twenty-four hours of birth. They rely on a sense of smell, being distressed and disoriented if the nest is cleaned. If the mother's teats are washed kittens may be unable to find their usual feeding positions.

Day 7
Birth weight has doubled. Female kittens are sometimes bigger and heavier than their brothers. The kittens can push themselves along with swimming movements, but cannot raise their bodies from the ground and cannot retract their claws. They can raise their heads. Eyes are beginning to open.

Week 4
The kittens can move with their bellies raised from the ground and can sit. Eyes are open and blue, ears are pricked. They can see, and can judge distances. Their milk teeth are erupting, first the incisors, then the canines. The kittens dislike being separated from their siblings and do not leave the nest voluntarily, but when together they are starting to be playful. The mother may now move the litter to a new nest.

Family life

Week 5
The kittens are now very mobile and very playful, but still stay together most of the time. They are starting to groom themselves and one another. They can balance standing on their hind legs and are starting to 'box' in this position. They can arch their backs, pounce, make sidestepping mock attacks and chase moving objects.

Week 6
The kittens are now chasing one another and can leap horizontally. They are eating some solid food taken from their mother's dish and use their litter tray.

Notes on the development of your kitten

Week 7

They are now feeding from their own dish. They use their litter tray for playing in as well as for its proper purpose. They greet one another and their mother, signalling with their tails, touching noses and rubbing against one another.

Week 8

Play-fighting is now intense. The kittens are playing with toys and are possessive about them. If the mother brings them prey they will eat it after they have practised throwing, pawing and biting it.

Exploring and learning

Month 3

The kittens are now weaned and their mother no longer allows them to suck. They are learning to behave correctly – and cautiously – towards adults. The mother may come into a false oestrus (pro-oestrus), followed by full oestrus a few weeks later. The young cats are allowed outdoors. They should be vaccinated against the most common viral diseases.

Month 4

In the open, the cats are establishing the ranges within which they will do their hunting. The more food they are given by humans the smaller their ranges will be. The kittens can extricate themselves from difficulty but may be frightened to descend from high places.

Month 5

The cats are scent-marking their ranges. If their mother is pregnant again they will be leaving home now and looking after themselves. They are starting to shed their milk teeth.

Month 6

Although the family may still be together, the young cats can now live independently. They remain friendly with one another, but each goes its own way. Cats of either sex can be neutered at this age.

Month 7

The cats are still exploring. New experiences constantly occur and they must decide how to deal with them. They will experiment with hunting techniques and try new prey species.

Month 8

They are young adults, but not sexually mature.

Coming of age

Month 10

Tabitha came into oestrus. Females mature sexually at five months to a year. Males secrete androgens at four months but cannot mate till twelve-fifteen months. Tabitha was made pregnant by Mr Tibbs.

Month 12

Tabitha gave birth to her own litter of five kittens.

Notes on the development of your kitten

Further reading

Allaby, Michael and Crawford, Peter, *The Curious Cat,* Michael Joseph Ltd, London, 1982. In 1978-79 the BBC Natural History Unit sponsored and filmed a study of a family of farm cats. The film, directed by Peter Crawford, was shown in *The World About Us* series. This book describes the project, adding many details the film could not include and supplying a general background to the life of those cats which have little or no contact with humans.

Beadle, Muriel, *The Cat,* Simon and Schuster, New York, 1977, Collins and Harvill Press, London, 1977. A comprehensive and very authoritative account of the origin and spread of the domestic cat, its biology and behaviour, and its relationship to humans, including its use in scientific research.

Greene, David, *Incredible Cats,* Methuen, London 1984. The author is an experimental psychologist with much experience of child development. His book contains many anecdotes but is especially good in its descriptions of communications — vocal and body language — used by cats, dreaming, and ways in which local variations in fur markings can be used to trace migrations of cats, and their human companions, across continents.

Kirk, Mildred, *The Everlasting Cat,* Faber and Faber, London, 1977. The history of the relationship between cats and humans and the hold cats have always had over our imagination, leading sometimes to the infliction of appalling suffering on cats, and at others to their deification.

McFarland, David (editor), *The Oxford Companion to Animal Behaviour,* Oxford University Press, Oxford, 1981. Not a book about cats in particular, but one that contains a wealth of information about evolution and animal behaviour in general, with clear, non-technical explanations of the techniques used to study animals and the terms scientists use.

Necker, Claire, *The Natural History of Cats,* Delta Books, New York, 1977. A zoologist with a special love of cats, Claire Necker divides her book into two parts dealing first with the history of cats and their relationship with humans, and then with their biology. Much of the book is an anthology of writing about cats. It contains a great deal of useful information in easily accessible form.

Sayer, Angela, *Encyclopedia of the Cat,* Octopus Books, London, 1984. With the technical help of a veterinarian, Angela Sayer devotes much of her book to the care of cats, with a useful chart to help diagnose ailments. She also describes cat breeds and provides information on showing cats.

Sillar, F.C. and Meyler, R.M., *Cats Ancient and Modern,* Studio Vista, London, 1966. A brief discussion of the history of the cat in art and literature leading to an anthology of prose, verse and pictures about cats, collected from all over the world and from classical to modern times.

Tabor, Roger, *The Wildlife of the Domestic Cat,* Arrow Books, London, 1983. The author is a biologist who describes the physiology, behaviour and social organization of cats clearly and simply, illustrating his account with many anecdotes.

Index

AUTHORS' NOTE

Jane Burton and Kim Taylor would like to acknowledge their thanks to:

Fred Thorne, for entrusting us with his 'special' cats including Snorkel (and after whom Fred the kitten was named); Lesley Russell (on whose birthday Snorkel's kittens were born), for help looking after all the rescue cats including Snorkel's family; Evelyn Smart, for her hard work in organizing the rescue work for Cranleigh Animal Rescue Centre; David Stadden BVSc, MRCVS, for veterinary advice and help with the radiograph of kitten embryos; Desmond Thackeray, for encouragement with the electronics; Weycolour Limited, Godalming, for once again processing our film with care and courtesy; and to Catherine Ahern, for allowing Mr Tibbs to come and stay with us.

We are also grateful to the following people who have kindly provided homes for the NINE LIVES cats: Linda Gamlin and David Burnie (Ferdinand); Susan and Mark Scovell (Barley); Percy Remnant (Lucifer); Margaret and Nigel Mercer, Simon, Stephen and Andrew (Septimus and Fred); Rita Brenton (Ginger); and Jane Witty (Fergus).

ACKNOWLEDGEMENTS

Editorial Director	Ian Jackson
Art Director	Nick Eddison
Designer	Gill Della Casa
Copy Editor and Proof Reader	Jocelyn Selson
Indexer	Michael Allaby
Production	Bob Towell

Eddison/Sadd Editions would like to acknowledge with grateful thanks the co-operation received from Bruce Coleman Limited during the production of this book.